Free-Space Optical Communication Systems for Next Generation Networks

Abhijeet Upadhya • Vivek K. Dwivedi •
Ghanshyam Singh

Free-Space Optical Communication Systems for Next Generation Networks

Springer

Abhijeet Upadhya
Jaypee Institute of Information Technology
Noida, Uttar Pradesh, India

Vivek K. Dwivedi
Jaypee Institute of Information Technology
Noida, Uttar Pradesh, India

Ghanshyam Singh
University of Johannesburg
Johannesburg, South Africa

ISBN 978-3-031-74804-2 ISBN 978-3-031-74805-9 (eBook)
https://doi.org/10.1007/978-3-031-74805-9

This Springer imprint is published by the registered company Springer Nature Switzerland AG
The registered company address is: Gewerbestrasse 11, 6330 Cham, Switzerland

Preface

The book provides gainful insights in the field of free-space optical (FSO) communication which is considered as the next frontier for future-generation broadband wireless networks. The FSO systems have been proposed as a solution for the last mile connectivity problem because of their distinguishable features such as an extremely high available bandwidth, low cost as well as ease of deployment, license-free spectrum, and improved channel security. This form of communication provides a very high-data-rate outdoor point-to-point communication link with low latency. Besides, FSO systems also offer other advantages such as reduced latency and enhanced security. FSO communications provide high data rates in Gbits/s ranges through the atmosphere for distance from a few hundreds of meters to a few kilometers. FSO technology is employed in a range of applications, such as communication between buildings, data centers, and security systems. Nevertheless, the performance of FSO systems is dependent on weather conditions and can be disrupted when there's rain, fog, or other environmental elements present. A clear line-of-sight between the transmitter and receiver is essential for optimal operation. FSO technology seems a promising solution to the connection of end-users to the service providers or to other existing networks. Significant factors which affect FSO links are atmospheric turbulence and pointing errors. Importantly, the loss in FSO systems is more due to high operating frequencies (smaller wavelengths). This suggests that simultaneous use of radio frequency (RF) systems such as microwave and millimeter wave (mmWave) links can overcome these limitations due to their relatively smaller frequency of operation. However, the RF links suffer heavily due to environmental conditions and are unable to provide very long-range and high-speed communication. Therefore, the FSO/RF hybrid system model where both RF and FSO systems coexist is a potential solution for ensuring the reliability of the link in different weather conditions. This is because links are affected differently depending on the weather conditions of both RF and FSO channels. An FSO system is used as the primary network, whereas an RF system acts as a secondary backup network in the FSO/RF hybrid system. This hybrid system can be used for even very long-distance free-space communication systems such as deep-space communication, aircraft-to-aircraft, inter-satellite, satellite-to-earth,

earth-to-satellite, satellite-to-aircraft, aircraft-to-satellite, aircraft-to-ground, ship-to-ship, and ship-to-infrastructure links with very high reliability. FSO links include chip-to-chip communication, indoor infrared (IR) or visible light communication (VLC), inter-building communication, and free-space laser communications including airborne, space borne, and deep space missions. Yet another implementation for combining RF and FSO systems is the so-called mixed RF/FSO relaying systems, where the advantages of both forms of communication systems are utilized. This form of cooperative relaying ensures coverage enhancement along with increased throughput. The book provides a detailed description of mixed radio RF/FSO relaying techniques as backhaul networks and analyzes its performance considering several key system parameters that directly impact its execution.

The study of mixed RF/FSO cooperative relaying becomes intriguing due to the involvement of two different forms of communication systems. The RF link is severely affected by small-scale fading and large-scale path loss including shadowing. On the other hand, the FSO link operates under the constraints of atmospheric turbulence and pointing errors. Apart from this, the combined RF/FSO systems are generally operated in the vicinity of the other communication systems and their coexistence induced unwanted signaling from the neighboring channels. Therefore, the practical implementations of the mixed RF/FSO relaying technology are interference limited and comprehensive analysis is required. The present book is proposed to cover all these factors simultaneously to determine the impact of important parameters on the performance of mixed RF/FSO relaying systems. The book presents the various generalized channel models that can be adopted to model RF and FSO link statistics. Further, it presents the modeling of amplify-and-forward (AF) and decode-and-forward (DF) forms of cooperative relaying schemes. The mixed systems containing two networks of RF and FSO links are used to make inter-networking with increased link reliability. The connectivity through the relay is an important feature of the proposed system. The most important benefit of this combined system is the availability of a link at any worst environmental condition and can work well in all the weather conditions. The most significant parameters to be considered for network selection in mixed RF/FSO systems are environment conditions and required communication distance. In addition to this, co-channel interference (CCI) is a major deteriorating factor that can severely impact the reliability and data throughput offered by the overall mixed RF/FSO backhaul networks which is discussed in the book in detail. Moreover, the book will enable the readers to understand the various mitigation techniques that can be utilized in mixed RF/FSO relaying in order to improve the overall user experience.

The major driving force for cutting-edge technology is the combination of endless thirst for knowledge, unequal intuitive intelligence, and limitless ambition for a better life. Due to rapid urbanization, the increasing number of users and growing demand for high-speed internet connectivity/data rate attract the FSO communication market. The FSO communication market is an emerging technology that uses light to transmit data wirelessly between two points. FSO communication technology can be implemented in scenarios such as networks between the unmanned aircraft systems (UAS) and satellites as well as various other applications, including last mile access, healthcare, airborne application, disaster recovery, storage area networks, data transmission, defense, and security, are one of cutting-edge next-generation wireless communications network. Currently, the FSO communication network market is driven by the rising demand for fast data transmission, growing use of FSO within 5G/B5G/6G networks, low latency, and the need for secure communication channels. As data-intensive applications become more commonplace and there is an increasing requirement for fast internet connections, there is an increasing demand for ultrafast transmission technologies such as FSO, which offers numerous advantages over conventional wired and wireless communication networks, such as increased bandwidth, decreased latency, and enhanced security. FSO is increasingly being employed in 5G networks to provide high-speed backhaul connectivity between base stations, which is also expected to fuel the market for FSO communications services. In addition, secure communication networks are necessary, and FSO technology offers high levels of security—making it a suitable option for emerging applications. The healthcare sector is one industry that stands to benefit greatly from FSO technology as it offers faster, more reliable connectivity for medical equipment. Overall, the future looks bright for the growing FSO communication market. Thus, the FSO technology is increasingly finding applications in cutting-edge fields such as healthcare and the military, giving the FSO communication market tremendous growth potential. FSO systems are expected to experience improved performance and dependability due to advances in FSO technology, such as adaptive optics and beam steering. As FSO provides a speedy and cost-effective backhaul solution, rising FSO adoption in 5G networks is predicted to fuel the growth of the FSO communication market.

At present, technical details with mathematical models for implementing FSO communication networks has been elaborated in this book. The unique features of this book are as follows.

- Introduction to FSO communication networks which has been proposed as an important technology for future generation wireless network systems.
- Performance analysis and wireless channel modeling schemes of mixed RF/FSO relaying systems.
- The impact of interference on the operation of RF/FSO backhaul networks.
- Comprehensive overview of several FSO communication schemes to enhance the performance of next generation networks.
- Understanding key factors on the FSO link that determine the overall operation such as turbulence, pointing errors and path loss.
- This book is intended for senior undergraduate and graduate students, scientists, and researchers working in academia and industries in the area of electrical/electronic and computer engineering particularly, in the field of wireless communication, computing, and networking technologies.

The topics covered in the book are concerned to the future generation wireless communication systems and its wireless channel modeling for the required error free data rate. In this way it will attract various stages of readers as given below:

- The book will be useful to researchers in academia and industry in conceptualizing the development of next generation wireless communication networks modeling. The study presented in the book has the capability to help the network engineers to reduce the initial cost of implementation of FSO systems based on the intriguing results derived in the book.
- Present an ideal reference that is not only constructive to researchers in academia and industry to learn more in this evolving field but also inspires future generation mobile communication researchers toward further innovation.
- For the under-graduate and post-graduate students, it will assist in learning a new technology.
- Overview of wireless communication evolutions as well as channel estimation for FSO-based future generation communication systems.
- The detailed mathematical analysis presented in the book will be beneficial for the students/scholars to follow while conducting investigation of various wireless communication systems.

This book comprises 10 chapters. The Chap. 1 provides a comprehensive review of the state-of-the-art FSO communication in detail along with the various factors that should be considered while designing and analyzing FSO communication systems. The chapter also thoroughly converses about the means by which mixed RF/FSO cooperative relaying strategy can be implemented.

Chapter 2 discusses the FSO systems which provide major advantages of large bandwidth, quick installation, and inherent security. The complete mathematical formulation has been performed on the optical signal fading using the α-μ statistical distribution. To elaborate more about the importance of cooperative relaying strategy, a combination of radio and FSO systems has been explored. The numerical results about system performance based on the derived metrics are meticulously illustrated and validated through the Monte-Carlo simulation results.

In Chap. 3, a hybrid RF and FSO that is mixed RF/FSO cooperative relaying systems are presented which is generally assumed to be immune to interference. Therefore, the co-channel interference (CCI) is analyzed in order to comprehend the performance of mixed RF/FSO relaying systems employing α-μ model for atmospheric turbulence along with pointing error on the FSO link. The results suggest that the mixed RF/FSO relaying systems are prone to interference.

Chapter 4 presents the investigation of a multi-user hybrid system with RF and FSO technologies in an AF relay network, considering asymmetric channel fading conditions. The analysis assumes Nakagami-m fading for RF channels and double generalized Gamma (D-GG) fading for the wireless optical channel, incorporating the impact of pointing errors. Through this system model, exact expressions are derived for the outage probability and average BER. Additionally, the outage characteristics are explored under high signal-to-noise ratio (SNR) conditions to ascertain the coding gain and diversity order afforded by the system. Theoretical findings are substantiated through Monte Carlo simulations to validate their accuracy.

Chapter 5 analyzes a UAV/FSO relaying system, incorporating key features of UAVs in the modeling process. The DF relaying protocol is used to investigate the impact of various UAV factors such as probability of LoS, terrain, large- and small-scale fading along with the optical turbulence and pointing error on the FSO link.

Chapter 6 explores a system where the RF link is extended using a high-speed optical wireless link. To increase the diversity on the FSO link, the Gamma-Gamma turbulence statistical model is employed. The RF link statistics for both intended and interfering signals are modeled using the Nakagami-m distribution, with an exact analysis presented for any value of the fading parameter m. This mathematical model provides an accurate framework for studying the impact of various parameters on system performance.

In Chap. 7, we present a non-orthogonal multiple access (NOMA) scheme to evaluate the performance of a mixed RF/FSO cooperative relaying system. Further, relay node within this hybrid RF/FSO framework functions under the DF protocol is presented. The RF links are modeled using the Nakagami-m distribution, which effectively represents the small-scale fading in RF communication channels. For the FSO links, the effects of atmospheric turbulence are described using the Malaga distribution, a comprehensive model that accounts for a broad range of turbulence conditions. The FSO link model used in this study also considers the impact of pointing errors and includes two types of optical demodulation techniques: Intensity Modulation/Direct Detection (IM/DD) and coherent demodulation. Additionally, the numerical results have been thoroughly validated through the Monte Carlo simulations.

Chapter 8 offers an overview of how machine learning models can be utilized in future wireless communication systems. Further, various existing contributions for the implementation of machine learning algorithms in the field of optical systems are explored and then the future research paths to use deep learning models for enhancement of FSO technology are suggested.

Chapter 9 provides various problems that are inherent in the modeling of the optical wireless channel. The research attempts to use the high dimension feature of the deep learning models to represent the complex varying FSO channel.

Finally, in Chap. 10, the channel estimation, which is crucial task in the next-generation wireless communication systems, is accurately determined between the transmitter and receiver, which significantly impacts the performance of various communication functions. Recently, deep learning models, particularly convolutional neural networks (CNNs), have brought about significant advancements in wireless communication systems. Additionally, long short-term memory (LSTM) models offer a compelling approach to processing even larger sets of input data, making them well suited for channel estimation tasks. This chapter delves into the modeling of optical wireless channels using both CNN and LSTM models, exploring their potential to enhance the accuracy and reliability of channel estimation in wireless communication systems.

In summary, the book provides a unified view of the state-of-the-art mixed RF/FSO next-generation communication networks, which should be accessible to a readership with basic knowledge about electromagnetic and wireless communication technology with signal processing. The readership may find the rich set of references in each of the chapters very useful. The authors have performed a good job by providing a concise summary of all the chapters at the preface of the book. We would strongly recommend the book to graduate students and researchers and engineers working or intending to work in the area of imaging and sensing application different area of applications. This book distinguishes itself from the existing prosperous literature of FSO and mixed RF/FSO technologies with emerging applications.

The authors are indebted to a numerous colleague for the valuable suggestions during the entire period of manuscript preparation. The first and second authors are thankful to Vice-chancellor, Jaypee Institute of Information technology, Noida, India, for providing all the needed support to complete the work. The first author conveys deep respect to his parents who have always blessed him with immense love. The first author dedicates all his efforts to his wife and son, who have constantly made him happy and blissful during the completion of the book. The second author conveys the greatest gratitude to his parents whose tremendous blessings and endless sacrifice always encouraged during the writing of the manuscript as well as carrier growth. They have shown immense patience and supported in every possible way. Words alone can never express the gratitude to them. The second author also extends warm thanks to his wife and son, their innocent smiles and chats always bring cheerfulness even in the odd time. We would also like to thank publishers at Springer, in particular Charles B. Glaser, Brian P. Halm, Kaitlyn Bonomo, and R. M. Bakiyalakshmi for their helpful guidance and encouragement during the creation of this book. The authors would not justify their work without showing the gratitude to their family members who have always been the source of strength to tirelessly work to accomplish the assignment. In the last but not least, We thank the ultimate source of energy of every particle in the universe, the Almighty, for giving the enough energy and strength to complete the work. All praise and gratitude belong to him.

Noida, Uttar Pradesh, India — Abhijeet Upadhya
Noida, Uttar Pradesh, India — Vivek K. Dwivedi
Johannesburg, South Africa — Ghanshyam Singh

Contents

About the Authors

Abhijeet Upadhya completed his PhD at JIIT, Noida, India, in 2020. He completed his Master of Technology and Bachelor of Technology at Dr. A.P.J. Abdul Kalam Technical University (AKTU), (formerly known as Uttar Pradesh Technical University (UPTU)), UP, India, in 2014 and 2011, respectively. Currently, he is serving as an assistant professor in the Department of Electronics and Communication Engineering at JIIT, Noida, UP, India. To date, he has published a number of quality research papers in SCI-indexed journals and continues to explore the vast area of Communication Engineering. He has also published one patent, which is under review for being granted under the Indian Patent Act, 1970. He constantly provides support as a reviewer to various prestigious journals. His research interests include wireless communication systems, cooperative relaying, and free space optical communication systems. Other areas of research include statistical characterization and modeling of wireless fading channels, security analysis, and performance evaluation of digital wireless communication systems subject to interference. He has also found interest in the field of deep learning and machine learning in wireless communication systems.

Vivek K. Dwivedi (Member, IEEE) received the Bachelor of Engineering degree from RGPV, Bhopal, India, in 2003; the Master of Engineering degree from the Birla Institute of Technology, Mesra, Ranchi, India, in 2006; and the PhD degree in electronics and communication engineering from the Jaypee University of Information Technology, Solan, India, in 2012. He was a senior visiting researcher in wireless communications with the University of Pretoria, South Africa, from 2014 to 2015. He is currently an associate professor with the Department of Electronics and Communication Engineering, Jaypee Institute of Information Technology, Noida, India. He has authored several research articles in refereed international journals and conferences. His primary research interests include machine learning, artificial intelligence, 5G, and optical wireless communication. He is serving as an Associate Editor of IEEE Access journal.

Ghanshyam Singh received his PhD degree in Electronics Engineering from the Indian Institute of Technology, Banaras Hindu University, Varanasi, India, in 2000. Currently, he is working as a full professor and director with the Centre for Smart Information and Communication Systems, Department of Electrical and Electronic Engineering Sciences, Auckland Park Kingsway Campus, University of Johannesburg, South Africa. He was associated with Central Electronics Engineering Research Institute, Pilani, and Institute for Plasma Research, Gandhinagar, India, respectively, where he was a research scientist. He had also worked as an assistant professor at Electronics and Communication Engineering Department, Nirma University of Science and Technology, Ahmedabad, India. He was a visiting researcher at the Seoul National University, Seoul, South Korea. He also worked as professor with the Department of Electronics and Communication Engineering, Jaypee University of Information Technology, Wakanaghat, Solan, India. He has more than 24 years teaching and research experience. His research and teaching interests include Millimeter/THz Wave Technologies and it's Applications in Communication and Imaging, Next-Generation Communication Systems (5G/6G)/Cognitive Radio/NOMA, Resource Allocation and Management, Interference Management, Applications of 5G/6G in Sustainable Smart City, Industry 4.0/5.0, Healthcare 4.0, Intelligent Transport System, Energy Management (IoE), and Digital Farming. He is an author/co-author of more than 370 scientific papers of the peer-reviewed journals and international conferences and 10 published books by Springer, Wiley, and CRC Press. He has supervised many PhD and MTech theses. He has worked as editor/associate editor and reviewer for several reputed journals and conferences.

List of Figures

Chapter 1
An Introduction to Free Space Optical Systems

1.1 Background

The advent of wireless technologies has seen tremendous growth in recent years. This has led to enormous bandwidth requirements due to the need to provide high data rate multimedia applications to users including video conferencing, telephony, and instant web browsing. The conventional radio frequency (RF) spectrum has therefore become one of the scarcest resources. Research arena has been forced to look beyond RF communications and investigate other forms of wireless communication systems. One of the major contenders that is being implemented for high bandwidth wireless transmission is the free space optical (FSO) communication. The complementing feature of the FSO technology is the key due to which it can be integrated with the existing RF systems. The frequency of operation for FSO communication includes infrared (IR) to visible light in FSO systems [1–10]. Notably, the optical spectrum is mostly unused for wireless data transmission. This is one of the many reasons why researchers are focusing on analyzing the optical domain for wireless transmission to obtain large throughput for future wireless communication systems. The present chapter presents an overview of the FSO communication system and provides details of significant contributions in the field of mixed RF/FSO relaying systems.

1.2 Free Space Optical Communication

1.2.1 Overview

There is no doubt that broadband services across the globe need very high data throughput. The wireless access technologies play an important role in determining the achievable data rate to the end user. There is no doubt that next generation

A. Upadhya et al., *Free-Space Optical Communication Systems for Next Generation Networks*, https://doi.org/10.1007/978-3-031-74805-9_1

wireless networks should be able to offer higher data throughput to support various broadband services. The current landscape of wireless access technologies encompasses a variety of options, including copper and coaxial cables, wireless Internet access, broadband RF/microwave, and optical fiber. However, certain limitations accompany these technologies. Copper and coaxial cables, as well as broadband RF/microwave systems, face challenges such as low data rates, expensive licensing, and congested spectrum. On the other hand, the implementation of optical fiber can be a complex task for network engineers, often involving cumbersome processes. In particular, the installation of optical fiber becomes economically unfeasible for rural areas, where digging the Earth for fiber deployment may not be a viable proposition. In light of these challenges, FSO communication systems have emerged as promising solutions to meet the data communication needs of various entities, including government, military, and commercial organizations [2–12]. FSO systems bring several advantageous features to the table, including high bandwidth, physical security, and the ability to facilitate wavelength reuse. These characteristics position FSO communication systems as potential game changers in addressing the shortcomings associated with other wireless access technologies. Furthermore, FSO systems operate within the unlicensed spectrum, eliminating any concerns related to licensing for the operator. A significant advantage of FSO systems lies in their swift deployment. The main advantage of FSO communication systems is the possibility to install an ad hoc communication link by configuring optical telescope toward an optical receiver aperture where the data transfer can take place wirelessly over the optical spectrum. These systems transmit data using an optical beam, employing low-power optical sources like lasers or light-emitting diodes (LEDs) within the visible range of the spectrum.

In contrast to the laborious process of excavating sidewalks for laying fiber links, FSO systems can be rapidly deployed, contingent upon the existence of a clear line-of-sight (LoS) link between the optical transmitter and the receiver aperture. This particular characteristic of FSO communication systems positions them as a viable solution for establishing networks in areas prone to natural disasters [7]. Moreover, the ease of deploying networks in remote locations, including borders, surveillance sites, challenging geographical landscapes, and ad hoc setups in deserts, underscores the versatility of FSO systems. FSO technology also finds practical application in establishing Satellite links, making it an excellent candidate for applications such as deep space probes and inter-satellite links, thanks to its compatibility with small-sized antennas [2, 13, 14]. The substantial bandwidth capabilities offered by FSO systems further contribute to its suitability for military applications. The adaptability of FSO technology to diverse scenarios makes it a valuable asset in addressing communication challenges in a variety of contexts.

Despite the myriad benefits offered by FSO systems, the wireless optical channel faces challenges due to atmospheric losses and varying weather conditions along the propagation path, as illustrated in Fig. 1.1. Key factors contributing to degradation in free space optical transmissions include rain, dust particles, fog, snow, path loss, turbulence-induced fading, and pointing errors [8–12]. Atmospheric turbulence, also known as scintillation, arises from rapid fluctuations in the intensity and phase

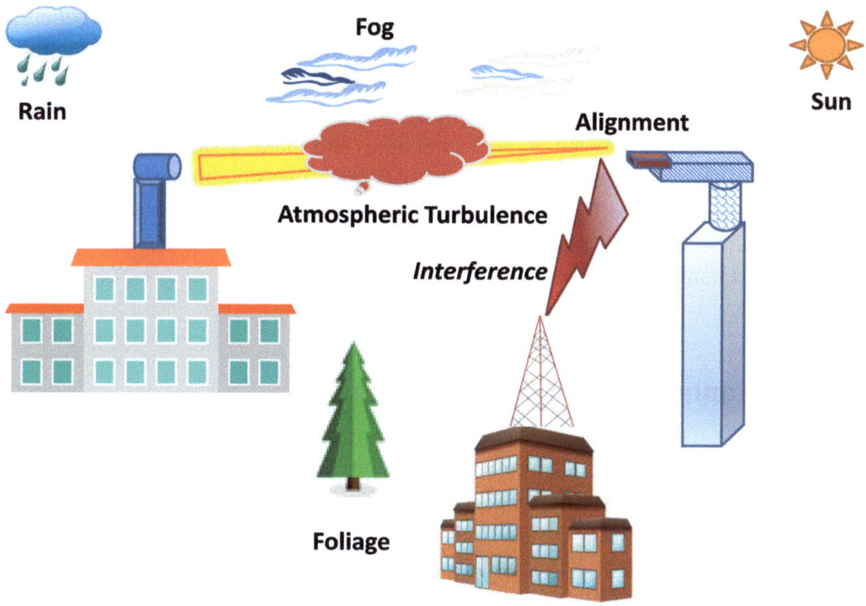

Fig. 1.1 Illustration of various factors that impede the signal transmission in the FSO communication systems

of the optical wave, triggered by changes in the refractive index of the medium [1]. This turbulence is a chaotic state induced by the flow of air, resulting from variations in temperature within the atmosphere. The atmospheric turbulent medium is characterized by numerous spherical regions or eddies with randomly varying diameters and different refractive indices. As the optical beam propagates through this turbulence channel, it encounters random temporal and spatial fluctuations in the refractive index medium, leading to significant fading effects. The interplay of these factors underscores the complexity of the atmospheric conditions impacting the wireless optical channel.

Pointing error, stemming from dynamic wind loads, building sway, and seismic tremors, leads to misalignment between transmitter and receiver apertures [3–5]. Additionally, the wireless optical network experiences path loss, determined by the Beer-Lambert law [4]. To address these challenges, various mitigation schemes have been proposed in the literature. Spatial diversity, forward error correction (FEC), automatic repeat request (ARQ), and dynamic thresholding have proven to be effective solutions [5, 15–20]. At the physical layer, aperture averaging is employed to overcome FSO system challenges. This technique involves increasing the size of the receiver aperture to average out fast fluctuations caused by small eddies [21]. Adaptive optics (AO), a feedback control system, is utilized in certain FSO systems to mitigate turbulence. This involves applying a pre-corrected beam to the transmitter before sending the signal into the turbulent channel. The pre-correction is achieved by applying the conjugate of the atmospheric turbulent channel to the beam. Optimization of FSO systems often involves selecting modulation schemes based on bandwidth and power efficiency. Key modulation schemes for optical communication include on-off keying (OOK) [22] and pulse position modulation (PPM) [23]. Other frequently adopted mitigation schemes in FSO communications include coding, jitter isolation and rejection, and background noise rejection. However, in last few years, the concept of cooperative communication has been used to enhance the performance of the FSO systems. In the following section, the importance of the cooperative relaying technique is discussed, serving as a mitigation scheme in the research work.

1.3 Advantages of FSO Communications

FSO systems function in the optical spectrum, typically using visible light or infrared spectrum. Unlike RF and millimeter-wave (mmWave) systems, which are limited by the congested and regulated electromagnetic spectrum, FSO has an advantage of a much larger bandwidth without licensing issues. The ideally enormous bandwidth available with FSO systems allows them to achieve data rates in the order of Tbps, making them ideal for applications requiring high throughput, such as fixed line connections, data centers, interactive video conferencing, and satellite communication. For implementing heterogeneous networks within urban scenario, the 6G systems will have to rely heavily on making FSO connectivity a feasible option for high data transfer. It is also important to integrate the FSO technology along with the existing RF and mmWave systems.

The effect of electromagnetic interference (EMI) is minimum in FSO systems when compared to traditional RF and mmWave systems. The conventional methods of communication are highly effected by the environments with high levels of electronic noise. This makes FSO a reliable option in industrial settings, military applications, and urban environments where EMI is prevalent. On the other hand, the highly directional nature of FSO communication reduces the likelihood of interception or jamming, a significant concern in RF and mmWave communications. Nowadays, when future wireless communication systems are making efforts to enhance security of wireless systems, it becomes imperative that FSO systems be implemented. Using the LoS feature to advantage, the secured transmission is possible with FSO systems. This feature is particularly beneficial for secure communications in military, government, and corporate networks where data security is paramount. Compared to RF and mmWave spectrum, there is no requirement of licensing in optical spectrum. This separates the FSO communication from all the other forms of communication strategies that are used for wireless communication. This makes it a cost-effective solution, particularly for emerging markets or regions where obtaining spectrum licenses may be challenging or expensive.

The time to deploy the FSO communication is very small. They are the main contenders for establishing ad hoc communication networks in remote areas and in adverse climatic situations. This facility is not evident in RF or any other contenders of wireless communication systems. They do not require extensive trenching or cabling, making them ideal for rapidly establishing communication links in disaster recovery scenarios, temporary events, or remote locations. FSO links can be easily scaled and redeployed as network demands change. This contrasts with RF and mmWave systems, where changes often require new hardware or significant reconfiguration. Additionally, the optical spectrum is less congested, allowing for easier network expansion without interference issues.

Compared with traditional optical fiber systems, the FSO systems have advantages of mobility and reusablity, cost effectiveness, ease of installation, and less dependency on weather conditions. The optical fiber installation is very cumbersome work and needs a lot of time and money. In situations that need quicker installation of network, optical fibers are generally not selected as an option. The FSO system hardware can be mobile in nature and can be taken from one place to another in case of any foliage. The cost effectiveness of FSO systems is a result of aforementioned advantages over optical fibers. The FSO communication links have been traditionally deployed to connect satellites with satellites and to connect satellite with ground stations as shown in Fig. 1.2. FSO can provide high-capacity links between satellites in low Earth orbit (LEO) and geostationary Earth orbit (GEO). These links are critical for creating a global satellite network capable of supporting real-time data transmission, Earth observation, and global Internet coverage. The connection between LEO satellites with mobile phones through FSO links is a more convenient way of cellular system where the base stations are

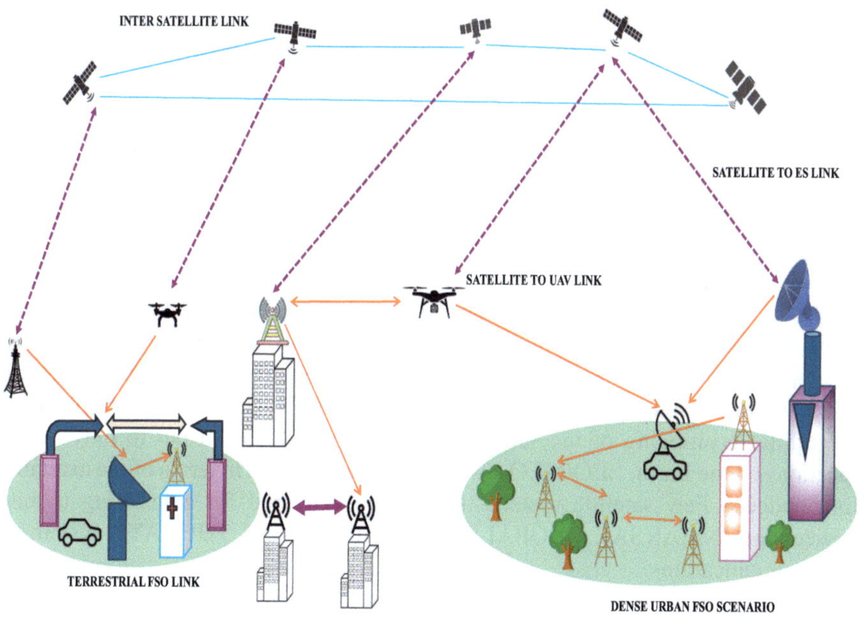

Fig. 1.2 FSO communication links to connect satellite to terrestrial networks and to establish inter-satellite links

the LEO satellites. Since they are nearer to the Earth, multiple LEO satellites are installed in the space, which function as routers for connecting mobile receivers across the borders. These forms of mobile satellite systems (MSS) also play an integral part to connect remote areas wirelessly. This is generally not possible through traditional form of communication.

FSO systems can establish robust communication links between satellites and ground stations. The high data rates and low latency of FSO make it an ideal solution for downloading large volumes of data, such as high-resolution images or videos, from satellites to the Earth. FSO is also used to connect terrestrial networks, particularly in areas where traditional fiber-optic cables are impractical to install due to geographical or environmental constraints. For example, FSO can bridge the "last mile" gap in urban areas, connect offshore platforms to onshore networks, or provide communication links in mountainous regions. The satellites can also be connected to unmanned aerial vehicles (UAVs) for better connectivity to far locations such as desert or oceans.

Hybrid mobile satellite systems (MSS) are the systems in which dual mode connection is provided using both satellite and mobile phones. The smooth functioning of MSS depends upon the interworking of terrestrial cellular systems and the space segment driven by the satellite systems. Deployment of an hybrid MSS provides the benefits of both these segments by exploiting the individual and coordinated efficiency between the two systems. Therefore, designing a hybrid communication system needs proper understanding of the separate systems as well as their combined technological aspects. The basic consideration that one has to make is about the two-fold limitation that exists in any MSS. Satellite link degradation restricts the highest possible efficiency in terms of the network optimization based on large separation between the transmitter and the receiver. Secondly, the requirements for evolution of the handheld mobile terminal with smaller size and cost also restrict some major aspects since it is not possible to exactly imitate the terrestrial mobile phones for an MSS. Price of services and the licensing problems are of the main concerns for the MSS operators. Prices generally mark the degree to which these services may reach to the common user. Licensing issues always seem to have put a hold on some objectives, although the principal reason that we have licensing is to minimize the chaos within the market. The issue can be minimized by transmitting on the optical carrier frequency using the FSO technology. LEO/MEO constellation provides the same capacity to all geographical areas irrespective to population density which results in inefficient utilization of total system capacity since oceans and deserts do not require connectivity. A GEO system may optimize such a scenario by planning the orbital positioning of the satellite in such a way to reach out to only desire locations on the Earth. LEO/MEO systems have advantages in terms of the available link budgets. Few satellite systems that have been used for connecting mobile devices are INMARSAT, GLOBALSTAR, IRIDIUM, and THURAYA system of telecommunications.

1.4 Cooperative Relaying Technology

Various implementations of multiple transmitter and receiver structures for wireless signaling have garnered attention in the literature [24–29]. Broadly, these multiple-input multiple-output (MIMO) systems can be categorized into the following three groups: The first category is the spatial multiplexing scheme, where multiple transmitters and receivers are utilized to transmit the data signal, achieving multiplexing gains. In such cases, multiple paths can be created between the transmitter and the receiver over which multiple streams of information can be sent and received. When the channel conditions are favorable, a spatial multiplexing scheme can be deployed to harness the possibility of transmitting more data over the same communication channel. However, when the channel conditions are not favorable, it is required to ensure that the transmitted data is received. This leads to the second category of MIMO implementation which is the spatial diversity scheme, where redundant data is transmitted from multiple transmitters to enhance the reliability of the wireless link. Various combining schemes are required to collect the signal at the receiver. Selection combining is scheme where only the signal with the largest signal-to-noise ratio (SNR) is processed, while in maximal ratio combining (MRC) scheme each signal is weighted based on its received signal strength [30]. The third category of MIMO schemes involves multiple transceiver systems that employ beamforming techniques, steering the transmitter antenna beam toward the desired target. It is worth noting that in all these schemes, the multiplexing and/or diversity gains come at the cost of increased hardware requirements.

To overcome the problem of increased hardware requirements applicable in the MIMO systems, cooperative relaying systems are used, which leverage existing devices to create diversity in communication systems. In a cooperative relaying strategy, communication between two nodes occurs with the assistance of the hardware of intermediate nodes. This mechanism can provide spatial diversity/multiplexing gains based on channel conditions, without increasing hardware requirements, as the intermediate nodes are typically the existing mobile user terminals in the communication system [31]. Since these multiple transmitters/receivers are distributed across several devices, these virtual MIMO systems are referred to as cooperative relaying schemes. Beyond offering diversity gains, such relaying systems have the potential to increase the coverage area of the network [10]. The main types of relaying systems found in the literature include:

1.4.1 Amplify-and-Forward Relay

The basis of the amplify-and-forward (AF) protocol is such that the information signal from the source node initially amplifies the analog signal and subsequently transmits the amplified version to the destination node. Notably, no decoding operation is performed in this process, leading to the characterization of these relays as non-regenerative relays. Moreover, AF relaying systems can be further categorized into fixed gain and variable gain networks. In fixed gain relays, the relay gain is set at the outset of data transmission [6, 32]. Conversely, in variable gain schemes, the relay gains depend on the channel state of the source-to-relay link [9, 33, 34]. The distinction between fixed gain and variable gain relays adds flexibility to AF relaying systems, allowing for adaptation to varying channel conditions during data transmission.

1.4.2 Decode-and-Forward Relay

In decode-and-forward (DF) relaying systems, the relay takes on the task of explicitly decoding the message transmitted by the source. Following the decoding process, the relay generates a new coded message and forwards it to the destination node [35, 36]. This relaying topology is referred to as regenerative relaying. There are typically three variants of the DF relay strategy. The first variant is the basic DF scheme, where the relay decodes the message in phase-I and forwards the new coded message in phase-II if the decoding in phase-I is successful. The second variant is the selection DF, where the source node is allowed to retransmit the original message itself in phase II if the relay cannot successfully decode in phase-I [31]. The third category of DF relaying is demodulate-and-forward (DeF) relaying. In this approach, if the relay cannot decode the message, it has the option to demodulate the message on a symbol-to-symbol basis. These variations provide a spectrum of options for DF relaying systems, allowing for adaptability to different decoding and forwarding scenarios.

1.4.3 Coded Cooperation

Coded cooperation relaying schemes can be seen as a generalized version of the DF relay scheme. In these cooperative relaying schemes, more potent channel codes are employed to enhance the spectral efficiency of the overall system [37–39]. Within this scheme, distinct portions of the same message are transmitted in the two phases, ensuring that the source and relay nodes do not transmit the same version of the message twice. Specifically, the source transmits the coded version of the message in phase-I without the incremental redundancy, and this incremental redundancy is then transmitted in phase-II, by either the source or the relay node. This approach optimizes the use of channel coding to improve the overall performance and efficiency of the cooperative relaying system.

1.5 Asymmetric Mixed RF/FSO Relaying Schemes

Over the past decade, FSO technology has captivated the research community's interest due to its favorable characteristics. Faced with the scarcity of the RF spectrum, FSO communication emerges as an encouraging prospect for broadband wireless access, either complementing or serving as an alternative to RF solutions to address existing "last mile" network challenges [6–9, 32–36, 40–49]. However, RF systems inherently exhibit robustness to fading due to lower frequency of operation and are generally implemented at a lower cost, making them suitable for non-line-of-sight (NLOS) forms of communication. Both RF and FSO channels are affected by meteorological and natural conditions. In adverse channel conditions, RF links tend to outperform FSO links. Moreover, for mobile communication concerns, RF technology remains the preferred choice in extreme environment conditions. Nonetheless, due to features such as large available bandwidth in the optical domain, low latency, and quick deployment, FSO systems have been proposed as candidates for backhauling, effectively multiplexing several RF channels. This has led to a significant focus on mixed RF/FSO relaying systems among researchers.

The authors of [6] have proposed an asymmetric RF/FSO relaying strategy, where the channel between the source and the relay node is assumed to be an RF link, while the relay to destination link is observed to be an FSO link. This form of relaying is termed asymmetric as both hops of communication undergo different levels of fading. Various channel models are considered for representing the complex envelope of radio systems, including Rayleigh, Nakagami-m, Rician, $\eta - \mu$, generalized-K fading models. On the FSO link, turbulence (or irradiance fluctuations) is modeled based on the degree of fading, ranging from weak to moderate to strong. For weak turbulence, the log-normal fading distribution is adopted, while for strong turbulence, the negative exponential distribution is utilized. The Gamma-Gamma turbulence model is capable of approximating a moderate to strong scintillation environment.

1.6 FSO Technology as Backhaul Networks

The demand for high-capacity data traffic on cellular networks of more than 500 exabytes has led to the deployment of numerous small cell base stations to provide faster data rates with lower latency [7, 50]. The overall performance of the communication system relies heavily on the technology used for backhaul to connect these base stations. FSO networks offer very attractive feature of high bandwidth which can be used for multiplexing several RF signals, making the system useful for backhauling. In this context, mixed RF/FSO relaying systems have emerged as a significant candidate for backhaul solutions [7]. A schematic of the system model for mixed FSO/RF cooperative relaying systems is illustrated in Fig.1.3. Traditional backhauling solutions have primarily involved copper,

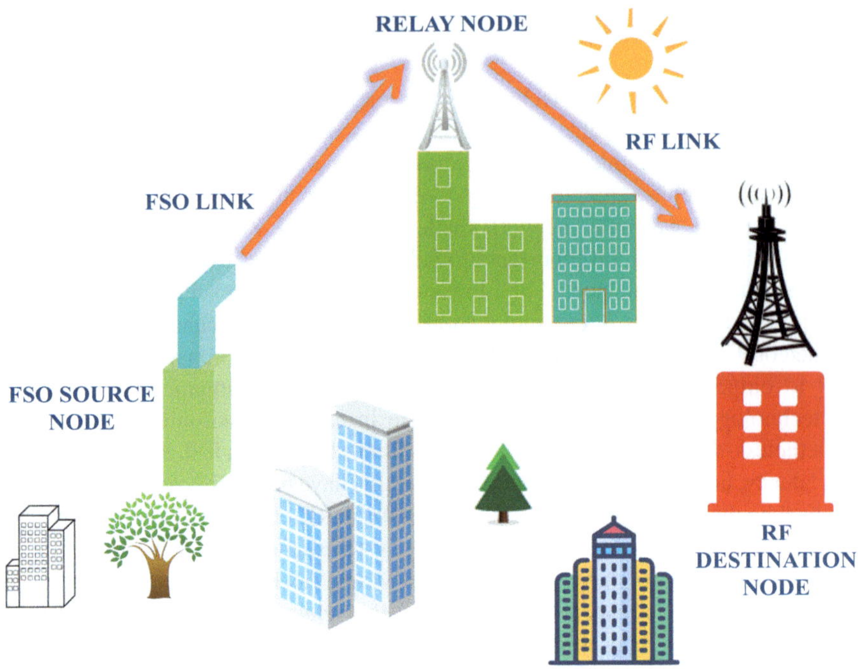

Fig. 1.3 A pictorial representation of mixed FSO/RF cooperative relaying systems

microwave, or optical fiber communications. Leased T1/E1 copper wire backhaul links have been utilized in second-generation (2G) and third-generation (3G) mobile communications, providing data rates of 1.544 Mbps for T1 and 2.048 Mbps for E1 carrier systems [51]. To meet the high data rate requirements of next generation mobile communications, multiple parallel connections of copper lines are often employed, resulting in increased costs. Optical fiber communication, while capable of delivering larger data rates, comes with a higher initial cost and deployment time. In locations where wired links are not deployable, wireless setups such as RF and FSO are preferred. RF backhaul networks operate in an NLoS environment but are subject to higher latency. In contrast, LoS FSO links can operate with very low latency compared to RF links but are susceptible to adverse weather conditions. The next generation of wireless systems introduces scenarios where the backhaul connections between two base stations involve a combination of NLoS and LoS links due to the presence of multiple intermediate base stations. In such situations, a combination of RF and FSO communication systems can enhance the overall user experience in terms of quality of service. Therefore, mixed RF/FSO cooperative relaying communication systems have the potential to provide a backhaul solution for next generation mobile networks.

1.7 FSO Channel Models

Modeling of the FSO links is the most involved portion for understanding the behavior of wireless medium. Various models exist in the literature that have been used to illustrate the physical medium. The statistical models correspond to various stengths of turbulence. The negative exponential irradiance probability density function (PDF) has been widely accepted to represent strong atmospheric turbulence [52]. The K-distribution has also been considered a good fit to model turbulence conditions [53]. Log-normal fading distribution can be used to represent the weak turbulence regime. A derived distribution from log-normal irradiance model is the log-normal K-distribution [54]. This is when the log-normal is joint with the K-distribution to enhance the modeling capability in order to representing the optical medium. Another similar model is the log-normal-Rician distribution [55]. Inverse-K distribution has also been extensively utilized for representing the fluctuation of optical signal as suggested in [56]. Among all these models, the major concern is that of the range of turbulence covered by the model. Most of these models have only capability to represent only a given range of turbulence. However, efforts have been made by the research community to derive statistical distributions which can represent a wider range for the strength of turbulence. On such model is the Gamma-Gamma turbulence model [57]. This distribution has the capability to represent moderate to strong turbulence condition. Very recently, turbulence models have been proposed that can represent the channel characteristics for weak-to-strong turbulence conditions. One such PDF is that of the double generalized Gamma (D-GG) turbulence model [58]. In this model, the selection of parameters can be varied to consider all regimes of the turbulence. The Malaga distribution is one other emerging PDF for representing the complex turbulence conditions on the FSO link for the complete range of scintillation [59].

1.7.1 FSO Relaying Systems

Cooperative relaying is an upcoming technique that has gained a lot of attention of researchers. It gives the system the benefit of using existing hardware to enhance the capacity and at the same time become energy efficient. Another benefit of relaying is that it increases the coverage of the overall system. One such cooperative technique has been proposed in the research work [6], where two hops are used to connect the source to the destination through an intermediate relay node. Since both the hops undergo different forms of communication, the relaying is said to be asymmetric. Specifically, the RF channel following Rayleigh fading has been considered for the first hop, while the second hop assumed FSO link over Gamma-Gamma fading model. The relay strategy adopted in [6] is AF relaying, making the overall system low cost with low complexity. Additionally, subcarrier intensity modulation (SIM) is considered as the type of modulation on the optical link. A closed-form expression for the outage probability has been derived and simulated to learn the behavior of the complete system since a lot of research has been carried by several researchers for availing the benefits of combining the RF and FSO systems to analyze the so-called asymmetric mixed RF/FSO cooperative relaying strategy. Although there is yet another way of combining the RF and FSO systems in a hybrid mode as suggested in [7] where one of the technologies can be utilized for communication at a given time. For mixed RF/FSO systems, the authors of [8] extend the research conducted in [6] to examine asymmetric relaying systems with pointing errors on the FSO link. Specifically, [8] assumes Rayleigh distributed fading for the RF link, while atmospheric turbulence affecting the optical wireless link is modeled as Gamma-Gamma distribution. Reference [8] provides precise closed-form expressions for the outage probability, bit error rate (BER), higher order fading statistics, and ergodic capacity. Furthermore, [9] showcases the RF channel with Nakagami-m m fading distribution, where the Rayleigh fading model is encompassed as a specific instance. Simultaneously, the turbulence affecting the FSO link is characterized using the Gamma-Gamma distribution. The study presented in [9] explores two variants of AF relaying strategy: fixed gain and variable gain relaying. Although the upper bound for analyzing the variable gain AF relay has been used, the results in the paper yield that variable gain relaying performs better than the fixed gain systems. The reason for this is the fact that based on the channel conditions on the first hop of communication, the relay modifies the amount of gain it provides before sending the signal to the destination in case of variable gain relays. After these initial results were presented, several researchers performed various forms of analysis on the pretext of mixed RF/FSO relaying. Few other works that remain important are [32–36]. In the study referenced as [32], the impact of hardware impairments is examined. Simultaneously, in the work presented in [33], the authors analyze the outage probability for variable gain mixed RF/ FSO AF relaying systems. In a comparable study, [34] explores a variable gain mixed RF/FSO AF relay network, focusing on the presence of outdated channel state information (CSI). Making another significant contribution, the research endeavors in [35, 36] showcase the performance of mixed DF relaying networks. Furthermore, in [40], the authors conduct an analysis of the performance of FSO systems under the influence of Malaga ($\mathcal{M}$) distribution. The generalization of the M model has been exploited in the paper to analyze outage probability, BER, and capacity of the network. Also, during the analysis, a PDF that can represent both forms of optical demodulation has been devised in the work. Building on this, the researchers in [41] introduce the partial relay selection (PRS) scheme, assuming the presence of CSI at the relay node for the mixed RF/FSO transmission scheme. Additionally, the impact of an eavesdropper on the mixed RF/FSO AF scheme is discussed in [42], where a jamming model is proposed to ensure secure information transmission. Exploring further, the study in [43] investigates the influence of higher order modulation on the RF link, coupled with the advantages offered by the FSO link.

Moreover, the authors in [44] analyze asymmetric RF/FSO relaying, where a millimeter-wave (mmWave) Rician distributed fading channel is assumed on the first hop, and the M distributed FSO channel is considered on the second hop. In a similar vein, [45] provides exact and asymptotic expressions for the outage probability and ergodic capacity in FSO/mmWave relaying systems. The cognitive radio perspective is explored in [46], where asymmetric RF/FSO relaying systems are analyzed. Here, the signal strength of users on the RF link is modeled as Nakagami-m fading distribution, and the scintillation on the optical wireless link is assumed to undergo D-GG fading. For cognitive radios, [47] delves into the analysis of MIMO RF/FSO DF relay networks, while the impact of aggregated hardware impairment on mixed RF/FSO relaying systems is examined by Balti et al. in [48]. Analyzing FSO systems with pointing errors over D-GG fading channels is the focus of [49], where exact and asymptotic closed-form expressions for BER, outage probability, and ergodic capacity of the mixed RF/FSO cooperative network are derived.

To enhance the overall performance of terrestrial FSO communication systems, multi-hop and bidirectional relaying are advocated, yielding various degrees of success and applications such as user diversity. In works like [60–63], the single-hop relay concept extends to multi-hop relaying schemes to improve energy efficiency and enhance coverage. In [60], the average BER and outage performances of multi-hop DF parallel relaying FSO communication systems are presented, considering the combined effects of path loss, pointing error, and atmospheric turbulence modeled as $\mathcal{M}$-distribution. Authors in [61] analyze the performance of serial relaying systems where the optical links undergo Gamma-Gamma turbulence model fading. Additionally, [62] investigates the performance of multi-hop relaying schemes with pointing errors. Extending the background in [62, 63] provides closed-form analytical expressions for BER and ergodic capacity for multi-hop FSO systems. The authors derive expressions for outage probability, BER, and ergodic capacity while considering the presence of outdated CSI for relay selection using PRS.

The wireless fading channel's spectral efficiency can be enhanced through the application of bidirectional relaying, known as two-way relaying (TWR) [64, 65]. This technique's application to FSO systems is explored in various studies, including [66–69]. In [66], the authors analyze the capacity of an all-optical TWR system using relay selection protocols and the moment generating function (MGF) approach. Meanwhile, [67] presents outage analysis of TWR systems. Recently, [68] performs a performance analysis of TWR FSO systems over D-GG turbulence channels with nonzero boresight pointing errors. To enhance the reliability of wireless fading channels, diversity schemes are employed. User diversity gains are achieved by optimally allocating resources to users with independently varying channels, improving overall system efficiency in terms of reliability and capacity. Opportunistic scheduling, where resources are allocated to users with the best channel conditions, is a notable approach, especially in the downlink. In FSO systems, multiuser diversity (MUD) schemes are incorporated in research works [69–74] for FSO transmission systems to enhance performance through user diversity.

In [69], the authors have investigated the performance of DF mixed RF/FSO TWR systems, improving the RF link using user diversity. Additionally, a power allocation scheme is presented in [69] to minimize the outage probability of the considered DF TWR system. Research work in [70] proposes a user scheduling algorithm, while [71] analyzes a hybrid FSO/RF point-to-multipoint system. The concept of effective capacity in terms of quality of service is explored in [72]. The work by Chen and Wand in [73] presents the outage performance of a fixed and variable gain AF relay network with diversity combining and selection. In a significant contribution, Yang et al. [74] consider the performance of the MUD-RF/FSO AF relaying system, where the user diversity aided RF link is modeled as $\eta - \mu$ distributed, and the turbulence on the FSO link is modeled using $\mathcal{M}$-distribution. The authors provide closed-form expressions for the BER, outage probability, and ergodic capacity of the relaying system.

1.7.2 *Interference in Mixed RF/FSO Systems*

The incorporation of radio systems in mixed RF/FSO relaying introduces vulnerability to interference from neighboring communication systems. Previous studies examining the performance of mixed RF/FSO relay systems have largely assumed interference-free transmission of information signals. However, introducing interference in asymmetric relaying can enhance the practicality of the system model. The impact of interference on the operation of mixed RF/FSO relaying systems has been explored in earlier works such as [75–80]. The paper [75] investigates the performance of mixed RF/FSO cooperative systems where RF transmission is used for one hop and FSO transmission is used for the other hop, in a dual hop configuration. It compares this to a direct RF link without the FSO hop. For the FSO link and RF links, it assumes D-GG turbulence with pointing error and Nakagami-m fading, respectively. It also considers the effect of co-channel interference (CCI) at the relay and destination. Under the assumptions of amplify-and-forward relaying and selection combining at the destination, it derives closed-form expressions for outage probability and bit error probability in terms of special functions. Its analysis provides a generalized framework that can recover several existing results, as the statistical models adopted for turbulence and fading are very general. Based on high SNR analysis, it also provides a diversity gain analysis and discusses potential bottlenecks in mixed RF/FSO systems. On the other hand, the authors of [76] investigated the performance of mixed radio/wireless optical relaying strategy where the relay is corrupted by both interference and noise. The RF link is modeled as Nakagami-m fading, while the FSO link experiences Gamma-Gamma fading from atmospheric turbulence including pointing errors. Monte Carlo simulations validate the analytical results. The expressions are simplified for an RF link limited by noise. Numerical results show an outage probability floor which should be considered in system design.

Additionally, in [77], the authors derive expressions for the ergodic capacity of the considered system, focusing on multi-aperture FSO/multiuser RF cooperative systems, where interference is taken into account. The impact of interference on the performance of mixed RF/FSO systems is analyzed in [78] over generalized fading channels. Furthermore, in [79], the authors conduct a performance analysis of multiuser mixed RF/FSO relaying systems in the presence of eavesdroppers. They present a power allocation scheme aimed at optimizing the outage performance of the system. In [80], exact and asymptotic closed-form expressions for outage probability, BER, and ergodic capacity are derived for an interference limited mixed RF/FSO system employing a PRS scheme. The interfering signals in [80] are modeled as Nakagami-m distributed, while the high-speed FSO link is modeled using a D-GG turbulence model. Interference has been considered in research work [81] when MUD has been adopted on the RF link, while both relay and destination nodes perform in the vicinity of interferers. Outage probability and BER analysis have been presented in the work. Moreover, the research contribution in [82] uses the D-GG fading model to represent the FSO channel, while the user diversity aided RF channel works in the presence of CCIs. The authors have derived the closed-form expressions for the outage probability and BER. A two-way mixed RF/FSO interference limited DF relaying strategy has been discussed in the work [83], where outage probability and BER analysis have been presented. In another work, relay selection has been presented to overcome the effects of interference on mixed RF/FSO cooperative relays [84]. The $\alpha - \mu$ turbulence model with nonzero boresight pointing error has been analyzed in the work [85] to quantify the impact of CCI in the vicinity of the relay node. The authors have derived closed-form expressions for the outage probability, BER, coding gain, diversity order, and capacity for the considered model.

1.7.3 Interference Management

FSO communication systems offer high data speeds and minimal delay, making them a promising for future wireless communication. However, these systems can be negatively impacted by various sources of interference, which can lead to performance degradation. Effectively managing interference is crucial to ensure reliable and robust FSO communication in dynamic and crowded wireless environments. One of the main obstacles in FSO communication is the presence of atmospheric turbulence, which has the potential to cause signal degradation and decrease the quality of the connection. Additionally, there is a significant risk of interference from various optical sources, including ambient light and competing FSO systems. These difficulties require the implementation of advanced interference management strategies to ensure the integrity of the signal and optimize the efficiency of the system. Effective management of interference plays a vital role in FSO communication systems to fully realize the potential of this technology in future wireless networks. By addressing atmospheric turbulence, optimizing the usage of different wavelengths, and implementing collaborative strategies for sharing the spectrum, FSO systems can achieve reliable and robust performance in various challenging environments. Following are few techniques that can be utilized to mitigate the impact of interference:

1.7.3.1 Adaptive Optics and Beamforming

Adaptive optics systems can be utilized to reduce the impact of atmospheric irradiance fluctuations. These systems dynamically adjust the optical path of incoming signals to compensate for distortions caused by turbulence, thereby improving the quality. Additionally, optical beamforming techniques assist in rejecting interference, allowing FSO systems to concentrate on desired signals while reducing unwanted sources.

1.7.3.2 Wavelength Management

Proper selection and management of transmission wavelengths are crucial in reducing interference. By carefully choosing wavelengths that minimize the impact of surrounding light and other optical sources, FSO systems can improve SNR values. This suggests that when the closeness of multiple FSO systems increases, it is possible to reduce the interference from the neighboring system by selecting a wavelength that is different and can be rejected using optical filers to mitigate the interference.

1.7.3.3 Collaborative Spectrum Sharing

In crowded wireless environments, it is possible to implement collaborative methods of sharing spectrum in order to coordinate FSO links and reduce interference between neighboring systems. By utilizing intelligent algorithms and dynamic spectrum access techniques, the efficient utilization of available spectrum resources can be facilitated.

1.8 Conclusion and Future Scopes

FSO indeed is a cutting-edge technology in the field of telecommunication. The laser beam used to transmit data over a wireless free space channel can be a scope for enhancing large data rate signals over optical spectrum with enhanced security. Recent advancements in FSO technology have significantly improved data rates, reliability, and overall performance. However, a lot of research is still pending and needs to be conducted before proper utilization of the potential of FSO communication. The FSO communication systems need to be integrated with unmanned aerial vehicles (UAV) for carrying out reliable information transfer at rapid data rates. On the other hand, deep learning algorithms can be applied to FSO systems integrated with satellite communication to improve signal processing, reducing the impact of atmospheric conditions on signal quality. In addition to this, quantum computing and artificial intelligence can be exploited to implement adaptive optics for overcoming the disadvantages of pointing error for enhancing the QoS.

The exponential growth of telecommunication systems has paved the path for various new technologies such as FSO for coping with the requirements of seamless connectivity at large data rates. The quantum leap in the requirement of data rates has made FSO systems a key contender for using optical spectrum to its fullest. The integration of FSO with existing radio technology will improve the overall network capacity, reduce latency, and contribute to the realization of a fully connected and intelligent communication ecosystem. The main advantage of using FSO systems is its feature of point-to-point link that will help in reducing the congestion in wireless networks. It is also expected that the quantum key distribution can be used even better for the security of the optical wireless systems. This will lead to high level of data integrity with confidentiality. Future wireless systems utilizing FSO technology should implement atmospheric compensation and adaptive optics techniques in order to elevate the pitfalls of the turbulence and pointing error. As we embark on the 6G era, FSO stands out as a key enabler for realizing the vision of a highly connected, intelligent, and secure communication infrastructure.

References

1. Z. Ghassemlooy, W. Popoola, S. Rajbhandari, Optical Wireless Communications: System and Channel Modeling with MATLAB (CRC Press, New York, 2012)
2. S. Arnon, J.R. Barry, G.K. Karagiannidis, R. Schober, M. Uysal, Advanced Optical Wireless Communication. (Cambridge University, Cambridge, 2012)
3. A.A. Farid, S. Hranilovic, Outage capacity optimization for free-space optical links with pointing errors. J. Lightwave Technol. **25**, 1702–1710 (2007)
4. H.G. Sandalidis, T.A. Tsiftsis, G.K. Karagiannidis, M. Uysal, BER performance of FSO links over strong atmospheric Turbulence channels with pointing errors. IEEE Commun. Lett. **12**, 44–46 (2008)
5. H. Kaushal, G. Kaddoum, Optical communication in space: challenges and mitigation techniques. IEEE Commun. Surv. Tutorials **19**, 57–96 (2017)
6. E. Lee, J. Park, D. Han, G. Yoon, Performance analysis of the asymmetric dual-hop relay transmission with mixed RF/FSO links. IEEE Photon. Technol. Lett. **23**(21), 1642–1644 (2011)
7. H. Dahrouj, A. Douik, F. Rayal, T.Y. Al-Naffouri, M.S. Alouini, Cost-effective Hybrid RF/FSO backhaul solution for next generation wireless systems. IEEE Wirel. Commun. **22**(5), 98–104 (2015)
8. I.S. Ansari, F. Yilmaz, M.S. Alouini, Impact of pointing errors on the performance of mixed RF/FSO dual-hop transmission systems. IEEE Wirel. Commun. Letters **2**(3), 351–354 (2013)
9. E. Zedini, I.S. Ansari, M.S. Alouini, Performance analysis of Mixed Nakagami-m and gamma-gamma dual-hop FSO transmission systems. IEEE Photonics J. **7**(1), 1–20 (2015)
10. M. Safari, M. Uysal, Relay-assisted free-space optical communication. IEEE Trans. Wirel. Commun. **7**(12), 5441–5449 (2008)
11. A. Jaiswal, M.R. Bhatnagar, V.K. Jain, Performance evaluation of space shift keying in free-space optical communication. IEEE/OSA J. Opt. Commun. Networking **9**(2), 149–160 (2017)
12. A. Jaiswal, M.R. Bhatnagar, V.K. Jain, Performance of optical space shift keying over gamma-gamma fading with pointing error. IEEE Photonics J. **9**(2), 1–16 (2017)
13. H. Kaushal, G. Kaddoum, Optical communication in space: challenges and mitigation techniques. IEEE Commun. Surv. Tutorials **19**(2), 57–96 (2017)

14. A.S. Hamza, J.S. Deogun, D.R. Alexander, Classification framework for free space optical communication links and systems. IEEE Commun. Surv. Tutorials **21**(2), 1346–1382 (2019)
15. R. Leoraj, J.A.V. Selvi, Comparative performance analysis of forward error correcting codes for free space optical communication, in *2016 International Conference on Emerging Trends in Engineering, Technology and Science (ICETETS)* (2016), pp. 1–6
16. Z. Abu Almaalie, Z. Ghassemlooy, A.A.S. Al-Rubaie, I.E. Lee, H. Le-Minh, Forward error correction with physical layer network coding in two-way relay free space optical links, in *2016 8th Computer Science and Electronic Engineering (CEEC)* (2016), pp. 1–5
17. J. Liu, X. Zhang, K. Blow, S. Fowler, Performance analysis of packet layer FEC codes and interleaving in FSO channels. IET Commun. **11**(13), 2042–2048 (2017)
18. K. Kiasaleh, Hybrid ARQ for FSO communications through turbulent atmosphere. IEEE Commun. Lett. **14**(9), 866–868 (2010)
19. V.V. Mai, A.T. Pham, Cross-layer designs and analysis of adaptive-rate transmission and ARQ for free-space optical communications. IEEE Photonics J. **8**(1), 1–15 (2016)
20. H.D. Le, V.V. Mai, C.T. Nguyen, A.T. Pham, Design and analysis of sliding window ARQ protocols with rate adaptation for burst transmission over FSO turbulence channels. IEEE/OSA J. Opt. Commun. Networking **11**(5), 151–163 (2019)
21. A. Viswanath, P. Gopal, V.K. Jain, S. Kar, Performance enhancement by aperture averaging in terrestrial and satellite free space optical links. IET Optoelectron. **10**(3), 111–117 (2016)
22. T. Joseph, H. Kaushal, V.K. Jain, S. Kar, Performance analysis of OOK modulation scheme with spatial diversity in atmospheric turbulence, in *2014 23rd Wireless and Optical Communication Conference (WOCC)* (2014), pp. 1–3
23. I.B. Djordjevic, B. Vasic, M.A. Neifeld, Multilevel coding in free-space optical MIMO transmission with Q-ary PPM over the atmospheric turbulence channel. IEEE Photon. Technol. Lett. **18**(14), 1491–1493 (2006)
24. P. Kumar, P. Pattanayak, K.M. Roy, Analysis of a New MIMO broadcast channel limited feedback scheduling algorithm with user grouping. Springer Wirel. Pers. Commun. **80**(3), 1079–1094 (2014)
25. Z. Wang, W. Zhong, S. Fu, C. Lin, Performance comparison of different modulation formats over free-space optical (FSO) turbulence links with space diversity reception technique. IEEE Photonics J. **1**(6), 277–285 (2009)
26. S. Jiang, G. Yang, Y. Wei, M. Bi, Y. Lu, X. Zhou, M. Hu, Q. Li, Performance analysis of space-diversity free-space optical links over exponentiated Weibull channels. IEEE Photon. Technol. Lett. **27**(21), 2250–2252 (2015)
27. B. Zhu, J. Cheng, L. Wu, A distance-dependent free-space optical cooperative communication system. IEEE Commun. Lett. **19**(6), 969–972 (2015)
28. M.T. Dabiri, M.J. Saber, S.M.S. Sadough, On the performance of multiplexing FSO MIMO Links in Log-Normal fading with pointing errors. IEEE/OSA J. Opt. Commun. Networking **9**(11), 974–983 (2017)
29. H. AlQuwaiee, I.S. Ansari, M. Alouini, On the maximum and minimum of double generalized gamma variates with applications to the performance of free-space optical communication systems. IEEE Trans. Veh. Technol. **65**(11), 8822–8831 (2016)
30. M.K. Simon, M.S. Alouini, Digital Communication over Fading Channels, 2nd ed. (Wiley, New York, 2005)
31. Y.-W. Peter Hong, W.-J. Huang, C.-C. Jay Kuo, Cooperative Communications and Networking: Technologies and System Design (Springer, New York, 2010)
32. E. Balti, M. Guizani, B. Hamdaoui, B. Khalfi, Mixed RF/FSO relaying systems with hardware impairments, in *2017 IEEE Global Communications Conference* (2017), pp. 1–6
33. M.I. Petkovic, Z. Trpovski, Exact outage probability analysis of the mixed RF/FSO system with variable-gain relays. IEEE Photonics J. **10**(6), 1–14 (2018)
34. M.I. Petkovic, G.T. Djordjevic, P.N. Ivanis, Partial relay selection with variable gain relays and outdated CSI in mixed RF/FSO system, in *2016 24th Telecommunications Forum (TELFOR)* (2016), pp. 1–4

35. M.R. Bhatnagar, Performance analysis of decode-and-forward relaying in gamma-gamma fading channels. IEEE Photon. Technol. Lett. **24**(7), 545–547 (2012)
36. P.K. Sharma, P. Garg, Bi-Directional Decode-XOR-Forward Relaying Over $\mathcal{M}$ -Distributed Free Space Optical Links. IEEE Photon. Technol. Lett. **26**(19), 1916–1919 (2014)
37. G. Yue, X. Wang, Z. Yang, A. Host-Madsen, Coding schemes for user cooperation in low-power regimes. IEEE Trans. Signal Process. **56**(5), 2035–2049 (2008)
38. M. Elfituri, W. Hamouda, A. Ghrayeb, A convolutional-based distributed coded cooperation scheme for relay channels. IEEE Trans. Veh. Technol. **58**(2), 655–669 (2009)
39. X. Bao, J.T. Li, Generalized adaptive network coded cooperation (GANCC): a unified framework for network coding and channel coding. IEEE Trans. Commun. **59**(11), 2934–2938 (2011)
40. I.S. Ansari, F. Yilmaz, M.S. Alouini, Performance Analysis of Free-Space Optical Links over $\mathcal{M}$álaga ($\mathcal{M}$) Turbulence Channels with Pointing Errors. IEEE Trans. Wirel. Commun. **15**(1), 91–102 (2016)
41. G. Djordjevic, M. Petkovic, A. Cvetkovic, G. Karagiannidis, Mixed RF/FSO relaying with outdated channel state information. IEEE J. Sel. Areas Commun. **33**(9), 1935–1948 (2015)
42. A.H. Abd El-Malek, A.M. Salhab, S.A. Zummo, M.S. Alouini, Security-reliability trade-off analysis for multiuser SIMO mixed RF/FSO relay networks with opportunistic user scheduling. IEEE Trans. Wirel. Commun. **15**(9), 5904–5918 (2016)
43. A. Touati, A. Abdaoui, F. Touati, M. Uysal, A. Bouallegue, On the effects of combined atmospheric fading and misalignment on the hybrid FSO/RF transmission. OSA J. Opt. Commun. Networking **8**(10), 715–725 (2016)
44. P.V. Trinh, T. Cong Thang, A.T. Pham, Mixed mmWave RF/FSO relaying systems over generalized fading channels with pointing errors. IEEE Photonics J. **9**(1), 1–14 (2017)
45. I. Trigui, N. Cherif, S. Affes, Relay-assisted mixed FSO-RF systems over Malaga-$\mathcal{M}$ and shadowed fading channels $\kappa - \mu$. IEEE Wireless Commun. Lett. **6**(5), 682–685 (2017)
46. H. Arezumand, H. Zamiri-Jafarian, E. Soleimani-Nasa, Outage and diversity analysis of underlay cognitive mixed RF-FSO cooperative systems. IEEE/OSA J. Opt. Commun. Networking **9**(10), 909–920 (2017)
47. N. Varshney, A.K. Jagannatham, Cognitive decode-and-forward MIMO-RF/FSO cooperative relay networks. IEEE Commun. Lett. **21**(4), 893–896 (2017)
48. E. Balti, M. Guizani, B. Hamdaoui, B. Khalfi, Aggregate hardware impairments over mixed RF/FSO relaying systems with outdated CSI. IEEE Trans. Commun. **66**(3), 1110–1123 (2018)
49. H. AlQuwaiee, I.S. Ansari, M.S. Alouini, On the performance of free-space optical communication systems over double generalized gamma channel. IEEE J. on Selec. Areas Commun. **33**(9), 1829–1840 (2015)
50. J.G. Andrews, What Will 5G Be?. IEEE JSAC **32**(6), 1065–1082 (2014)
51. Y. Li, M. Pioro, V. Angelakisi, Design of cellular backhaul topology using the FSO technology, in *Proceedings of the 2nd International Workshop Optical Wireless Communication Newcastle Upon Tyne, U.K* (2013), pp. 6–10
52. L.C. Andrews, R.L. Phillips, Laser Beam Propagation Through Random Media, 2nd edn. (SPIE Press, Bellingham, WA, 2005)
53. E. Jakeman, On the statistics of K-Distributed noise. J. Phys. Am. **13**, 31–48 (1980)
54. R.L. Phillips, L.C. Andrews, Measured statistics of laser light scattering in atmospheric turbulence. J. Opt. Soc. Am. **71**(12), 1440–144 (1981)
55. J.H. Churnside, S.F. Clifford, Log-Normal Rician probability density function of optical scintillations in the turbulent atmosphere. J. Opt. Soc. Am. **4**, 1923–1930 (1987)
56. L.C. Andrews, R.L. Phillips, I-K distribution as a universal propagation model of laser beams in atmospheric turbulence. J. Opt. Soc. Am. A2 **2**, 160–163 (1985)
57. M. Al-Habash, L.C. Andrews, R.L. Phillips, Mathematical model for the irradiance probability density function of a laser beam propagating through turbulent media. Opt. Eng. **40**(8), 1554–1562 (2001)

58. M.A. Kashani, M. Uysal, M. Kavehrad, A novel statistical channel model for turbulence-induced fading in free-space optical systems. IEEE/OSA J. Lightwave Technol. **33**(11), 2303–2312 (2015)
59. A.J. Navas, J.M.G. Balsells, J.F. Paris, A.P. Notario, A unifying statistical model for atmospheric optical scintillation, in *Numerical Simulations of Physical and Engineering Processes, Ch. 8* (2011)
60. P. Wang, R. Wang, L. Guo, T. Cao, Y. Yang, On the performances of relay-aided FSO system over $\mathcal{M}$ distribution with pointing errors in presence of various weather conditions. Opt. Commun. **367**, 59–67 (2016)
61. T.A. Tsiftsis, H.G. Sandalidis, G.K. Karagiannidis, N.C. Sagias, Multihop free-space optical communications over strong turbulence channels. IEEE Int. Conf. Commun. **6**, 2755–2759 (2006)
62. X. Tang, Z. Wang, Z. Xu, Z. Ghassemlooy, Multihop free-space optical communications over turbulence channels with pointing errors using heterodyne detection. J. Lightwave Technol. **32**(15), 2597–2604 (2014)
63. E. Zedini, M.S. Alouini, On the performance of multihop heterodyne FSO systems with pointing errors. IEEE Photonics J. **7**(2), 1–10 (2015)
64. D.S. Gurjar, P.K. Upadhyay, D.B. da Costa, R.T. de Sousa Jr., Beamforming in traffic-aware two-way relay systems with channel estimation error and feedback delay. IEEE Trans. Veh. Technol. **66**(10), 8807–8820 (2017)
65. P.K. Sharma, P.K. Upadhyay, Performance analysis of cooperative spectrum sharing with multiuser two-way relaying over fading channels. IEEE Trans. Veh. Technol. **66**(2), 1324–1333 (2017)
66. P. Puri, N.D. Chatzidiamantis, P. Garg, M. Aggarwal, G.K. Karagiannidis, Two-way relay selection in multiple relayed FSO networks. IEEE Wirel. Commun. Letters **4**(5), 485–488 (2015)
67. J. Park, C. Chae, G. Yoon, Amplify-and-forward two-way relaying system over free-space optics channels. J. Commun. Networks **19**(5), 481–492 (2017)
68. L. Yang, M.S. Alouini, I.S. Ansari, Asymptotic performance analysis of two-way relaying FSO networks with nonzero boresight pointing errors over double-generalized gamma fading channels. IEEE Trans. Veh. Technol. **67**(8), 7800–7805 (2018)
69. Y.F. Al-Eryani, A.M. Salhab, S.A. Zummo, M.S. Alouini, Two-way multiuser mixed RF/FSO relaying: performance analysis and power allocation. IEEE/OSA J. Opt. Commun. Networking **10**(4), 396–408 (2018)
70. J. Abouei, K.N. Plataniotis, Multiuser diversity scheduling in free-space optical communications. J. Lightwave Technol. **30**, 1351–1358 (2012)
71. L. Chen, W. Wang, C. Zhang, Multiuser diversity over parallel and hybrid FSO/RF links and its performance analysis. IEEE Photonics J. **8**(3), 1–9 (2016)
72. L. Chen, W. Wang, Effective capacity of MIMO free-space optical systems over gamma gamma turbulence channels. Opt. Commun. Elsevier **382**, 450–454 (2017)
73. L. Chen, W. Wang, Multi-diversity combining and selection for relay-assisted mixed RF/FSO system. Opt. Commun. **405**, 1–7 (2017)
74. L. Yang, M.O. Hasna, I.S. Ansari, Unified performance analysis for multiuser mixed η— μ and $\mathcal{M}$—distribution dual-hop RF/FSO systems. IEEE Trans. Commun. **65**(8), 3601–3613 (2017)
75. E. Soleimani-Nasab, M. Uysal, Generalized performance analysis of mixed RF/FSO cooperative systems. IEEE Trans. Wirel. Commun. **15**(1), 714–727 (2016)
76. M.I. Petkovic, A.M. Cvetkovic, G.T. Djordjevic, G.K. Karagiannidis, Outage performance of the mixed RF/FSO relaying channel in the presence of interference. Wirel. Pers. Commun. **96**(2), 2999–3014 (2017)
77. N. Cherif, I. Trigui, S. Affes, On the performance analysis of mixed multi-aperture FSO/multiuser RF relay systems with interference, in *18th IEEE International Workshop on Signal Processing Advances in Wireless Communications (SPAWC)* (2017), pp. 1–5

78. I. Trigui, N. Cherif, S. Affes, X. Wang, V. Leung, A. Stéphenne, Interference-limited mixed Málaga-$\mathcal{M}$ and generalized-K dual-hop FSO-RF systems, in *Proceedings IEEE Annual International Symposium Personnel, Indoor, and Mobile Radio Communication, Montreal, QC, Canada* (2017), pp. 1–6
79. A.H.A. El-Malek, A.M. Salhab, S.A. Zummo, M. Alouini, Effect of RF interference on the security-reliability trade-off analysis of multiuser mixed RF/FSO relay networks with power allocation. J. Lightwave Technol. **35**(9), 1490–1505 (2017)
80. E. Balti, M. Guizani, Mixed RF/FSO cooperative relaying systems with co-channel interference. IEEE Trans. Commun. **66**(9), 4014–4027 (2018)
81. A. Upadhya, V.K. Dwivedi, G. Singh, Multiuser diversity for mixed RF/FSO cooperative relaying in the presence of interference. Opt. Commun. **442**, 77–83 (2019)
82. A. Upadhya, V.K. Dwivedi, M.S. Alouini, Interference-limited mixed MUD-RF/FSO two-way cooperative networks over double generalized gamma turbulence channels. IEEE Commun. Lett. **442**, 1551–1555 (2019)
83. V.K. Tonk, A. Upadhya, P.K. Yadav, V.K. Dwivedi, Mixed MUD-RF/FSO two way decode and forward relaying networks in the presence of co-channel interference. Opt. Commun. **464**, 125415 (2020)
84. S.B. Bambiwal, A. Upadhya, R.S. Yaduvanshi, V.K. Pandey, Partial relay selection for combating the effects of co-channel interference in RF/FSO cooperative relaying. Opt. Commun. **475**, 126186 (2020)
85. A. Upadhya, V.K. Dwivedi, G.K. Karagiannidis, On the effect of interference and misalignment error in mixed RF/FSO systems over generalized fading channels. IEEE Trans. Commun. **68**(6), 3681–3695 (2020)

Chapter 2
Performance Analysis of Mixed RF/FSO Systems Over $\alpha - \mu$ Fading Channels

2.1 Introduction

This chapter provides an in-depth examination of mixed RF/FSO fixed gain relaying systems operating in the presence of atmospheric turbulence, characterized by the α–μ fading distribution. To elaborate more about the importance of cooperative relaying strategy, a combination of radio and FSO systems has been explored. Within the specified system, the RF hop of the AF system is presumed to undergo Rayleigh distributed fading, while the relay to destination link incorporates the effects of pointing errors on the optical wavelength. A detailed presentation of the PDF for the FSO link experiencing atmospheric turbulence modeled by α-μ is included. More specifically, the complex nature atmospheric turbulence on the wireless optical channel has been modeled using the new mathematical model. The investigation extends to the analysis of outage probability and BER for the considered system, presenting closed-form expressions articulated in terms of Meijer-G functions. Furthermore, an analysis of ergodic capacity for the cooperative relaying system is conducted, resulting in a closed-form expression represented in terms of bivariate Fox's H function. The chapter also presents the analysis of DF mixed RF/FSO relaying system in terms of outage probability. This comprehensive study sheds light on the influence of turbulence, pointing errors, the chosen optical demodulation scheme, and the binary modulation schemes employed within the RF/FSO relaying network. The numerical results about system performance based on the derived metrics are meticulously illustrated and validated through Monte Carlo simulation results. This multifaceted analysis contributes to a better understanding of the intricate dynamics affecting RF/FSO relaying networks in diverse environmental conditions.

A. Upadhya et al., *Free-Space Optical Communication Systems for Next Generation Networks*, https://doi.org/10.1007/978-3-031-74805-9_2

2.1.1 Background

The field of optical wireless communications (OWC) has captured the attention of researchers, primarily driven by the scarcity of available RF spectrum. To address the ever-growing demand for wireless data, researchers advocate for the implementation of FSO communication systems [1]. The optical domain's lack of licensing issues positions FSO systems to overcome the limitations of existing wireless technologies [2]. Despite the potential benefits offered by FSO technology, practical implementations face distinct challenges, including atmospheric turbulence and pointing errors [3]. Sending information signal using optical wavelength over free space is a complicated process and needs proper attention. The inhomogeneities within free space lead to variations in pressure and temperature, causing fluctuations in the refractive index of the medium. This, in turn, gives rise to atmospheric turbulence-induced fading, posing a significant challenge for FSO systems [3]. Simultaneously, pointing errors (or misalignment errors) arise due to discrepancies in the alignment between the transmitter telescope and the receiver's aperture, attributed to minor tremors and dynamic wind loads. These pointing errors play a crucial role in OWC [2].

Various strategies have been introduced in the literature to counteract signal degradation factors, including adaptive modulation and coding (AMC) [4], equalization and diversity [5], and cooperative relaying [2]. The wealth of bandwidth available in the optical domain positions FSO systems as promising contenders for the backhaul network of wireless communication systems. The mixed RF/FSO relaying, as proposed by [1], represents an instantiation of this concept and has been explored in numerous research works in the past including [6–11]. This system model adeptly leverages the advantages of both FSO and RF systems. Cooperative relaying techniques, such as those employed in mixed RF/FSO systems, are favored over diversity techniques due to their ability to combat fading and enhance coverage without escalating hardware requirements. Additionally, the selection of an accurate channel model to represent signal propagation mechanisms in the wireless optical domain is crucial. The present research work underscores the importance of utilizing generalized turbulence statistical models for this purpose.

2.1.2 Role of Accurate Channel Modeling

In the ever-changing world of wireless communication systems, the significance of precise channel modeling cannot be emphasized enough. Channel modeling forms the foundation for the design, optimization, and evaluation of wireless networks. The crucial role that accurate channel modeling plays in ensuring the effectiveness and dependability of wireless communication cannot be neglected before implementation of a practical system. An accurate representation of the wireless channel plays a crucial role in the effective design, optimization, and implementation of wireless communication systems. It provides engineers with the ability to anticipate, mitigate, and adapt to the challenges presented by the wireless channel, ultimately contributing to the establishment of strong and high-performing wireless networks. Following are some basic needs for modeling of wireless medium:

2.1.2.1 Understanding the Propagation of Signals

Accurate modeling of channels plays a crucial in comprehending the complexities of signal propagation. It enables engineers to simulate the effects of various environmental factors such as path loss, multipath fading, and interference, thereby offering insights into challenges presented by wireless channels. A thorough understanding of signal propagation dynamics is vital for the design of networks that can consistently deliver reliable performance in real-world scenarios.

2.1.2.2 Optimizing Link Budgets

Channel modeling serves as a key tool in optimizing link budgets. By accurately representing signal attenuation, noise, and interference, engineers can fine-tune power levels, modulation schemes, and antenna configurations to maximize the signal-to-noise ratio. This optimization is critical to establishing communication links that are reliable and possess sufficient margin for different deployment scenarios.

2.1.2.3 Predicting Performance and Designing Protocols

Accurate channel modeling facilitates the prediction of system performance under diverse conditions, aiding in the design and selection of communication protocols. Engineers can evaluate the effectiveness of modulation and coding schemes, as well as error correction techniques, to tailor them to the specific characteristics of the channel. This predictive capability ensures that wireless communication systems are resilient and capable of adapting to the challenges presented by various environments.

2.1.2.4 Adapting to Changing Conditions

Wireless channels are inherently dynamic, as they are subject to changes in the environment and user behavior. Accurate channel modeling enables the development of adaptive communication systems that can dynamically adjust their parameters in response to changing conditions. This adaptability is essential for maintaining optimal performance in the face of uncertainties.

2.1.3 The α-μ Statistical Model in Wireless Systems

Historically, atmospheric turbulence has been characterized by various statistical models, such as the negative exponential (NE) model for deep saturation conditions [12], the log-normal (LN) model for weak turbulence [13], and the K-distribution for strong turbulence conditions [14]. Recently, fading models like the Weibull distribution [15] and the D-GG distribution [16] have gained prominence for their ability to approximate existing models. Notably, the Gamma-Gamma turbulence model has received extensive recognition due to its capacity to represent moderate to strong turbulence conditions. Another model, the M distribution, is chosen as a more generalized irradiance fluctuation model [17].

In recent times, the α–μ fading model has gained acceptance for representing irradiance fluctuations in wireless optical signals [18]. This distribution accounts for the nonlinearities of the wireless fading channel, considering the modulus of various multipath components. The α–μ distribution accurately models channel fluctuations, with the α parameter representing nonlinearity and the μ parameter accounting for the number of multipath clusters.

The most essential aspect of designing and analyzing FSO systems is the modeling of the channel, which describes how the optical channel behaves between the sender and receiver. The $\alpha - \mu$ distribution has become a valuable tool for modeling the FSO channel in recent years. This distribution, which is an extension of the well-known Nakagami-m distribution, provides a versatile framework for capturing the fading characteristics of the optical channel. The parameters α and μ determine the shape and severity of the fading, allowing for a flexible representation of different atmospheric conditions. The $\alpha - \mu$ distribution is particularly suitable for FSO channels because it can model a wide range of fading scenarios, from weak to severe, which are commonly encountered in free space propagation. The distribution takes into account both multipath and scattering effects, resulting in a more realistic representation of the dynamics of the channel. This is crucial in FSO systems, where atmospheric turbulence, rain, and other environmental factors can cause significant fading and degradation of the signal. By utilizing the $\alpha - \mu$ distribution in FSO channel modeling, researchers and engineers can gain a better understanding of the statistical behavior of the optical channel. This understanding is fundamental for designing robust communication systems that can adapt to different environmental conditions. Additionally, accurate channel models facilitate the development of efficient modulation and coding schemes tailored to the specific characteristics of the FSO channel.

Fading channels are typically characterized by scattering effects in clusters of multipath components. The nonlinearity generated in the wireless medium, represented as the power of an exponent [18], is added as an exponent in the exponential relation of the Gamma density function. This arbitrary power offers better control in fitting experimentally collected data but increases mathematical complexity. The parameter μ can be used to model nonzero correlation between multipath component clusters if it is not restricted to being an integer [19].

The α–μ fading model has been adopted in the literature to model FSO links [20–22]. This model's deducement capability has been leveraged to approximate the envelope of the sum of Gamma-Gamma variates [20], model FSO MIMO channels [21], and simplify the analysis of ergodic capacity for Gamma-Gamma turbulence channels [22]. In this chapter, the mixed RF/FSO AF and DF relaying systems are analyzed, with the FSO link modeled as α–μ distribution under the influence of misalignment errors. The fading on the RF link is represented by the Rayleigh distribution.

2.2 Mixed RF/FSO AF Relaying System Model

Consider a dual hop mixed RF/FSO relaying system, where the source node S communicates with the destination node D through a relay node R. The system model assumes the absence of a direct link between nodes S and D, leading to communication occurring in two stages or hops. In the first hop, the AF relay node R receives an RF signal from the source node S. This relay node operates by converting the received RF signal into the optical domain and subsequently retransmits it to the destination node D. The system incorporates both forms of optical detection schemes: IM/DD and coherent detection. IM/DD, involving the detection of the light signal's intensity only, is known for its simplicity and cost-effectiveness [23–26]. On the other hand, coherent detection schemes, capable of overcoming the impact of thermal noise, have the potential to enhance the overall performance of FSO systems [27, 28]. The literature primarily considers heterodyne detection as the form of coherent detection in such systems.

The RF signal y_1 received at the relay node on the first hop of communication is given as

$$y_1 = a_1 x_1 + n_1 \tag{2.1}$$

where a_1 denotes the Rayleigh fading coefficient and x_1 is the RF signal transmitted from source node S. Moreover, n_1 denotes the additive white Gaussian noise (AWGN) component on the RF link with Power Spectral Density (PSD) N_{01}. Additionally, it is assumed that the relay node introduces a gain G and transmits the amplified version over the optical wireless link, which undergoes $\alpha - \mu$ turbulence. Assuming that the relay node employs SIM technique for modulation, the received optical signal at the destination node is expressed as [1, Eq. (3)].

$$y_2 = GI[1 + \eta(a_1 x_1 + n_1)] + n_2 \tag{2.2}$$

where I denotes the irradiance fluctuations. In (2.2), the constant η is the electrical-to-optical (E/O) conversion efficiency and n_2 is the AWGN component on the FSO link with PSD N_{02}. As per the presented model, the overall SNR random variable for the fixed gain AF relaying system is given as [1]

$$\gamma_{eq} = \frac{\gamma_1 \gamma_2}{\gamma_2 + c} \tag{2.3}$$

where the SNRs on both the hops are defined as $\gamma_1 = \frac{a_1^2 X_1}{N_{01}}$ and $\gamma_2 = \frac{\eta^2 I^2}{N_{02}}$ for IM/DD form of optical demodulation with X_1 denotes the average power of signal on the first hop of communication. However, for coherent demodulation, the relationship between irradiance power and SNR can be defined as $\gamma_2 = \frac{\eta I}{N_{02}}$. The value of constant c is given as $c = (G^2 N_{01})^{-1}$ [1]. Rayleigh distribution has been considered as the statistical PDF for representing the instantaneous SNR of random variable γ_1 on the RF link as per [29] as given below

$$f_{\gamma_1}(\gamma_1) = \frac{1}{\bar{\gamma}_1} \exp\left[-\frac{\gamma_1}{\bar{\gamma}_1} \right] \tag{2.4}$$

where $\bar{\gamma}_1$ denotes the average SNR for the $S - R$ link.

On the second hop of the communication, the atmospheric turbulence-induced irradiance I_a on FSO link has been modeled as $\alpha - \mu$ distribution with following PDF

$$f_{I_a}(I_a) = \frac{\alpha \mu I_a^{\alpha\mu - 1}}{\Gamma(\mu) \hat{I}_a^{\alpha\mu}} \exp\left[-\mu \left(\frac{I_a}{\hat{I}_a} \right)^{\alpha} \right] \tag{2.5}$$

In Eq. (2.5), $\hat{I}_a = \sqrt[\alpha]{\mathbb{E}[I_a^\alpha]}$ represents the α-root mean value of I_a [18], where $\mathbb{E}[.]$ denotes the expectation operator. Utilizing [18, Eq. (10)], the plot of the irradiance PDF is illustrated in Fig. 2.1. In this figure, the bar graph represents the statistics of the generated $\alpha - \mu$ random variates for $\alpha = 1.25$, $\mu = 1$. The close alignment of the generated $\alpha - \mu$ random variables around the theoretical PDF is evident in Fig. 2.1.

Additionally, by considering a Gaussian intensity profile with a waist of w at distance z, and incorporating independent identically distributed (IID) radial displacements of the Gaussian profile, the approximate PDF of the pointing (misalignment) error variable I_p on the FSO link is provided as per [30]

$$f_{I_p}(I_p) = \frac{\xi^2}{A_o^{\xi^2}} {I_p}^{\xi^2 - 1}, \quad 0 \leq I_p \leq A_o \tag{2.6}$$

where the $\xi = \frac{w_{eq}}{2\sigma_s}$ is the pointing error parameter which is defined as [30]. Here, the standard deviation of pointing error is denoted by σ_s, whereas w_{eq} represents equivalent beam radius. The role of ξ is to demonstrate the magnitude of pointing error such that its lower values correspond to larger pointing error and vice versa. The equivalent beam radius of the receiver can be defined as $w_{eq}^2 = \frac{w^2 \sqrt{(\pi)} \text{erf}(v)}{2v \exp(-v^2)}$, where $v = \frac{\sqrt{\pi} a}{\sqrt{2} w}$. The constant A_o is defined as the fraction of collected power at $a = 0$ and is given by $A_o = [\text{erf}(v)]^2$ [30] (Fig. 2.2).

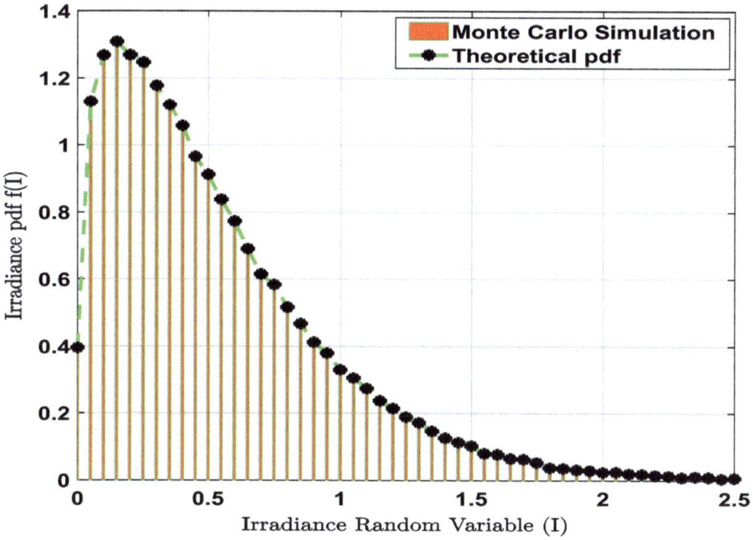

Fig. 2.1 Generation of $\alpha - \mu$ random variable

Fig. 2.2 Pointing error model

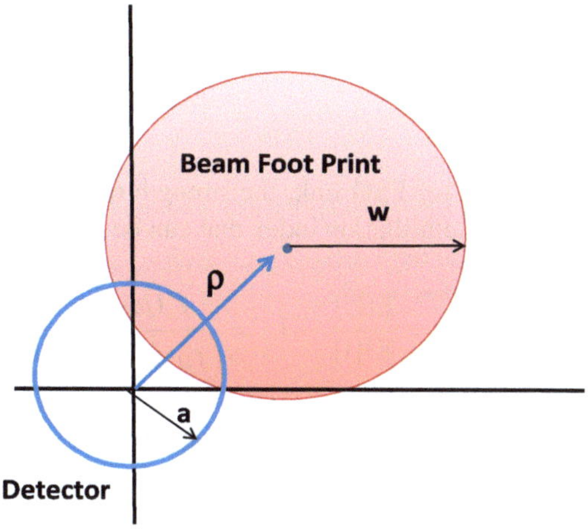

The overall FSO link statistical model includes the simultaneous impact of turbulence and pointing error. The analysis is generally conducted by considering that both the forms of fading are independent in nature. This suggests that the overall PDF random variable can be modeled using product of the random variables of the pointing error and turbulence. Therefore, the overall irradiance I in (2.2) can be expressed as $I = I_a I_p$. Additionally, assuming that the atmospheric turbulence and misalignment error variables are independent, the FSO channel statistics can be evaluated using

$$f(I) = \int f_{I/I_a}\left(\frac{I}{I_a}\right) f(I_a) dI_a \tag{2.7}$$

where the conditional probability $f_{I/I_a}\left(\frac{I}{I_a}\right)$ is given as $\frac{1}{I_a} f_{I_p}\left(\frac{I}{I_a}\right)$. Plugging the required values from (2.5) and (2.6) in (2.7), the required PDF can be further expressed in integral form as

$$f(I) = \int_{\frac{I}{A_o}}^{\infty} \frac{\xi^2}{A_o{}^{\xi^2}} \frac{\alpha \mu^\mu}{\hat{I}_a^{\alpha\mu} \Gamma(\mu)} I^{\xi^2 - 1} I_a^{(\alpha\mu - 1 - \xi^2)} \exp\Big[- \mu \Big(\frac{I_a}{\hat{I}_a}\Big)^\alpha \Big] dI_a \tag{2.8}$$

Moreover, integral in (2.8) can be evaluated using [31, Eq. (3.383.1)], and the overall channel state can be further presented as given below

$$f(I) = \frac{\xi^2}{\Gamma(\mu)} \Big(\frac{\mu}{A_o \hat{I}_a}\Big)^{\frac{\xi^2 + \alpha\mu - \alpha}{2}} W_{\frac{\alpha\mu - \xi^2 - \alpha}{2\alpha}, \frac{\xi^2 - \alpha\mu}{2\alpha}} \Big(\frac{\mu}{\hat{I}_a^\alpha}\Big) I^{\frac{\xi^2 + \alpha\mu - \alpha}{2} - 1}$$
$$\times \exp\Big[- \frac{\mu}{2} \Big(\frac{I}{\hat{I}_a A_o}\Big)^\alpha \Big] \tag{2.9}$$

where $W_{a,b}(c)$ represents the Whittaker function as defined in [31, Eq. (9.220.4)]. Utilizing the results provided in [18], the υth moment of the random variable I_a can be defined as follows:

$$\mathbb{E}^\upsilon[\mathrm{I}_a] = \frac{\Gamma(\mu + \frac{\upsilon}{\alpha})}{(\mu)^{\frac{\upsilon}{\alpha}} \Gamma(\mu)} \tag{2.10}$$

Generally, for FSO link, the strength of irradiance fluctuations can be represented using the scintillation index that can be formulated with the aid of (2.10) as

$$\sigma_I^2 = \frac{\mathbb{E}[\mathrm{I}^2]}{\mathbb{E}^2[\mathrm{I}]} - 1 \underset{\xi^2 \gg 1}{=} \frac{\Gamma(\mu)\Gamma(\mu + \frac{2}{\alpha})}{(\mu)^{\frac{1}{\alpha}} \big[\Gamma(\mu + \frac{1}{\alpha})\big]^2} \Big(\frac{\xi^2}{\xi^2 + 2}\Big) - 1 \tag{2.11}$$

In order to extract Rytov variance σ_1^2, the scintillation index σ_I^2 defined in (2.11), the obtained results can be compared with [32, Eq. (7, 9)]

$$\sigma_I^2 \simeq \begin{cases} \sigma_1^2 \left(1 + 0.5\sigma_1^2\right), & \sigma_1 << 1 \\ 1 + 0.86/\sigma_1^{4/5}, & \sigma_1 >> 1 \end{cases} \tag{2.12}$$

Nevertheless, the Rytov variance is defined by $\sigma_1^2 = 1.23 C_n^2 k^{\frac{7}{6}} L^{\frac{11}{6}}$, and λ denotes the signal wavelength, where C_n is the refraction structure parameter for a length of L and $k = \frac{2\pi}{\lambda}$ denotes the optical wave number. The higher required order of the PDF can be expressed using [18] as

$$\frac{\Gamma^2\left(\mu + \frac{1}{\alpha}\right)}{\Gamma(\mu)\,\Gamma\left(\mu + \frac{2}{\alpha}\right) - \Gamma^2\left(\mu + \frac{1}{\alpha}\right)} = \frac{\mathbb{E}^2[I_a]}{\mathbb{E}[{I_a}^2] - \mathbb{E}^2[I_a]} \tag{2.13}$$

and

$$\frac{\Gamma^2\left(\mu+\frac{2}{\alpha}\right)}{\Gamma(\mu)\,\Gamma\left(\mu+\frac{4}{\alpha}\right)-\Gamma^2\left(\mu+\frac{2}{\alpha}\right)}=\frac{\mathbb{E}^2[{I_a}^2]}{\mathbb{E}[{I_a}^4]-\mathbb{E}^2[{I_a}^2]} \tag{2.14}$$

The moment based estimators for the parameters α and μ of the distribution have been given as the transcendental equations in (2.13) and (2.14) in [18]. Therefore, by evaluating first, second, and forth moments of random variates of any arbitrary distribution, it is possible to generate $\alpha-\mu$ random variables that represent irradiance models such as K-distribution [33], Gamma-Gamma [1, 2], and log-normal distribution. Generally, mathematical software such as MATLAB and Mathematica can be used to solve the system of equations given in (2.13)–(2.14).

The derived PDF of the atmospheric turbulence needs to be represented in terms of SNR such that analysis can be conducted. If the IM/DD modulation scheme is implemented, the definition of electrical SNR μ_1 can be used for random variable transformation. Specifically, $\mu_1=\frac{(\eta\mathbb{E}[I])^2}{N_{02}}$ for IM/DD scheme. Performing the mathematical intermediate steps, the PDF of the $\alpha-\mu$ distribution can be expressed as follows [34]:

$$f(\gamma)=\frac{\xi^2}{2\Gamma(\mu)}\left(\frac{\mu}{\Omega_1 A_o\hat{I}_a}\right)^{\frac{\xi^2+\alpha\mu-\alpha}{2}} W_{\frac{\alpha\mu-\xi^2-\alpha}{2\alpha},\frac{\xi^2-\alpha\mu}{2\alpha}}\left(\frac{\mu}{\hat{I}_a^{\alpha}}\right)\gamma^{\frac{\xi^2+\alpha\mu-\alpha}{4}-1}\exp\left[-\mathrm{B}\gamma^{\frac{\alpha}{2}}\right] \tag{2.15}$$

where $\Omega_1=\frac{\sqrt{\mu_1}}{\mathbb{E}[\mathrm{I}]}$ and $\mathrm{B}=\frac{\mu}{2}(\Omega_1\hat{I}_a A_o)^{-\alpha}$. Similar to IM/DD, the irradiance on the FSO link can be derived for coherent detection using $\mu_2=\frac{(\eta\mathbb{E}[I])}{N_{02}}$ [3], where μ_2 represents average SNR on the FSO link for the case of coherent detection. The PDF can be obtained as [34]

$$f(\gamma)=\frac{\xi^2}{\Gamma(\mu)}\left(\frac{\mu}{\Omega_2 A_o\hat{I}_a}\right)^{\frac{\xi^2+\alpha\mu-\alpha}{2}} W_{\frac{\alpha\mu-\xi^2-\alpha}{2\alpha},\frac{\xi^2-\alpha\mu}{2\alpha}}\left(\frac{\mu}{\hat{I}_a^{\alpha}}\right)\gamma^{\frac{\xi^2+\alpha\mu-\alpha}{2}-1}\exp\left[-\mathrm{B}\gamma^{\alpha}\right] \tag{2.16}$$

It is now possible to combine the expressions in (2.15) and (2.16) to represent the unified PDF of the distribution as follows [34]:

$$f(\gamma)=\chi_1\gamma^{\frac{\xi^2+\alpha\mu-\alpha}{2r}-1}\exp\left[-\mathrm{B}\gamma^{\frac{\alpha}{r}}\right] \tag{2.17}$$

where $\chi_1=\frac{\xi^2}{r\Gamma(\mu)}\left(\frac{\mu}{\Omega_r A_o\hat{I}_a}\right)^{\frac{\xi^2+\alpha\mu-\alpha}{2}} W_{\frac{\alpha\mu-\xi^2-\alpha}{2\alpha},\frac{\xi^2-\alpha\mu}{2\alpha}}\left(\frac{\mu}{\hat{I}_a^{\alpha}}\right)$. In the combined PDF expression, as earlier stated, the constant $r=2$ represents the IM/DD detection technique, while $r = 1$ denotes the coherent detection of optical signal with $\Omega_r = \frac{(\mu_r)^{\frac{1}{r}}}{\mathbb{E}[I]}$ [34].

2.2.1 *Choice of Demodulation Scheme*

IM/DD depends on the modulation of the optical signal's intensity directly for intensity modulation. The intensity of the optical signal is varied in accordance with the modulating signal. The same needs to be detected by the demodulator. Direct detection means that the receiver measures the intensity of the received optical signal without attempting to retrieve the phase information. In comparison to coherent systems, IM/DD systems are simpler and more cost-effective. They do not require complex optical components such as local oscillators or phase-locked loops. IM/DD systems are less affected by changes in wavelength, making them suitable for situations where precise wavelength control is difficult. However, compared to coherent systems, IM/DD systems may have limited reach, especially when fiber impairments are present. The direct detection of intensity in IM/DD systems can result in information loss, particularly in the presence of channel impairments. Heterodyne detection involves mixing the received signal with a local oscillator at the receiver, allowing for the recovery of both amplitude and phase information. Coherent detection is a form of heterodyne detection. Heterodyne detection enables the application of advanced modulation formats, which offer high spectral efficiency and data rates. Additionally, heterodyne detection typically supports longer communication distances and is more tolerant to fiber impairments. The choice between IM/DD and heterodyne detection depends on the specific requirements of the application. IM/DD is often preferred for short-distance and cost-sensitive applications, while heterodyne detection is suitable for long-distance and high-capacity scenarios.

2.3 Performance Analysis of Mixed RF/FSO AF Relaying System Model

The statistical measures established in the preceding section will be employed in this segment to derive exact and asymptotic formulations for the outage probability, BER, and ergodic capacity of the envisioned system. The results derived in the subsection cover almost all the important signal degradation factors that a mixed RF/FSO relay system will undergo.

2.3.1 Outage Probability

The outage probability is characterized as the likelihood of the end-to-end system encountering signal outage when the instantaneous SNR descends below a preestablished threshold denoted as γ_{th}. Generally speaking, it provides a measure of the reliability of the communication system. It is an even more important characteristic for the wireless channel. The computation of the outage probability is articulated as follows [1]:

$$P_{out} = \int_0^\infty Pr[\gamma_{eq} < \gamma_{th}|\gamma_2] f_{\gamma_2}(\gamma) d\gamma = \int_0^\infty F_{\gamma_1}\left(\frac{\gamma_2 + c}{\gamma_2}\gamma_{th}\right) f_{\gamma_2}(\gamma_2) d\gamma_2 \tag{2.18}$$

2.3.1.1 Exact Analysis

To determine the probability of service interruption due to outage for the considered dual hop system, we can express the necessary CDF of the Rayleigh distributed SNR random variable on the radio (RF) link as $F(\gamma) = [1 - \exp(\frac{-\gamma}{\bar{\gamma_1}})]$. This CDF can be substituted in (2.18) as

$$P_{out} = 1 - \chi_1 \int_0^\infty x^{\frac{\xi^2+\alpha\mu-\alpha}{2r}-1} \exp\left[-\frac{c\gamma_{th}}{\bar{\gamma_1}x}\right] \exp\left[-Bx^{\frac{\alpha}{r}}\right] dx \tag{2.19}$$

The exponential functions involved in (2.19) can be represented in terms of Meijer-G function with the aid of [35, Eq. (07.34.03.0046.01)] as $\exp(-\frac{\gamma}{\bar{\gamma_1}}) = G_{0,1}^{1,0}\left[\frac{\gamma}{\bar{\gamma_1}}\middle| \begin{matrix} - \\ 0 \end{matrix}\right]$, to further express the integral as

$$P_{out} = 1 - \chi_1 \int_0^\infty x^{\frac{\xi^2+\alpha\mu-\alpha}{2r}-1} G_{0,1}^{1,0}\left[\frac{c\gamma_{th}}{\bar{\gamma_1}x}\middle| \begin{matrix} - \\ 0 \end{matrix}\right] G_{0,1}^{1,0}\left[Bx^{\frac{\alpha}{r}}\middle| \begin{matrix} - \\ 0 \end{matrix}\right] dx \tag{2.20}$$

The above equation can be transformed by changing the order of the first Meijer-G function using [35, Eq. (07.34.03.0228.01)]

$$P_{out} = 1 - \chi_1 \int_0^\infty x^{\frac{\xi^2+\alpha\mu-\alpha}{2r}-1} G_{1,0}^{0,1}\left[\frac{\bar{\gamma_1}x}{c\gamma_{th}}\middle| \begin{matrix} 1 \\ - \end{matrix}\right] G_{0,1}^{1,0}\left[Bx^{\frac{\alpha}{r}}\middle| \begin{matrix} - \\ 0 \end{matrix}\right] dx \tag{2.21}$$

where $G_{p,q}^{m,n}\left[Arg\middle| \begin{matrix} - \\ - \end{matrix}\right]$represents the Meijer-G function defined in [31, Eq. (9.301)]. The closed-form expression for the outage probability can be obtained as per the identity [35, Eq. (07.34.21.0013.01)], which is expressed as [34]

$$P_{out} = 1 - \chi_1 \frac{(g)^{\frac{1}{2}} (h)^{-\frac{(\xi^2+\alpha\mu-\alpha)}{2r}-\frac{1}{2}}}{(2\pi)^{\frac{g+h-2}{2}}} \exp\left[-\frac{\gamma_{th}}{\bar{\gamma_1}}\right] \left(\frac{c\gamma_{th}}{\bar{\gamma_1}}\right)^{\frac{\xi^2+\alpha\mu-\alpha}{2r}} G_{0,g+h}^{g+h,0}\left[B_1\middle| \begin{matrix} - \\ d_1 \end{matrix}\right] \tag{2.22}$$

In (2.22), the ratio $\frac{h}{g} = \frac{\alpha}{r}$ defines the order of G function, whereas $g + h$ terms in d_1 are the parameters of Meijer-G as given below

$$\mathrm{d}_1 = \frac{0}{g}, \ldots, \frac{0 + (g-1)}{g}, \frac{0 - \frac{(\xi^2+\alpha\mu-\alpha)}{2r}}{h}, \ldots, \frac{(h-1) - \frac{(\xi^2+\alpha\mu-\alpha)}{2r}}{h}$$

Moreover, the argument of Meijer-G function is derived as $B_1 = \left(\frac{\mathrm{B}}{g}\right)^g \left(\frac{c\gamma_{th}}{\bar{\gamma}_1 h}\right)^h$.

2.3.1.2 Asymptotic Analysis

Although the result in Eq. (2.22) provides exact analysis of outage probability, it is generally required to derive the simplified asymptotic approximate expression for the outage probability in order to illustrate the behavior of considered system from system designer's perspective. Generally, the system designer is concerned with the effect of various parameters at high SNR values, and the asymptotic approximation can provide with quick and important insights into system functionalities. In order to represent the results into elementary functions, the Meijer-G function can be expanded for small arguments as per [36]

$$\mathrm{G}_{p,q}^{m,n}\left[Arg\middle| \begin{smallmatrix} a_k \\ b_k \end{smallmatrix}\right] \simeq \sum_{i=1}^{m} \frac{\prod_{k=1}^{m} \Gamma(b_k - b_i) \prod_{k=1}^{n} \Gamma(1 - a_k + b_i)}{\prod_{k=n+1}^{p} \Gamma(a_k - b_i) \prod_{k=m+1}^{q} \Gamma(1 - b_k + b_i)} (Arg)^{b_i} \tag{2.23}$$

Applying the above-mentioned identity in (2.22), the exact outage probability expression can be converted in terms of elementary functions as given below:

$$P_{out} \underset{\bar{\mu}_r \gg 1}{\approxeq} 1 - \chi_1 \frac{(g)^{\frac{1}{2}} (h)^{-\frac{(\xi^2+\alpha\mu-\alpha)}{2r} - \frac{1}{2}}}{(2\pi)^{\frac{g+h-2}{2}}} \exp\left[-\frac{\gamma_{th}}{\bar{\gamma}_1}\right] \left(\frac{c\gamma_{th}}{\bar{\gamma}_1}\right)^{\frac{\xi^2+\alpha\mu-\alpha}{2r}}$$
$$\times \sum_{j=1}^{g+h} \left[\left(\frac{g}{\mathrm{B}}\right)^g \left(\frac{\bar{\gamma}_1 h}{c\gamma_{th}}\right)^h\right]^{-\mathrm{d}_{1,j}} \prod_{i=1}^{g+h} \Gamma(\mathrm{d}_{1,i} - \mathrm{d}_{1,j}) \tag{2.24}$$

In (2.24), $\mathrm{d}_{1,j}$ denotes the jth term in d_1. Since B in (2.24) is inversely proportional to electrical SNR, therefore, for higher average SNR, the asymptotic expression is dominated by $\min[\mathrm{d_j}]$ as per the location of poles in the contour integral.

2.3.2 *Bit Error Rate*

The closed-form expression of BER has been derived corresponding to both kinds of optical detection techniques for the considered dual hop relaying system. Notably,

BER analysis over wireless fading channels has been performed by utilizing the MGF based approach or using the characteristic function (CF) as given in [37–39]. However, a new and useful relation has been proposed in [40], using which BER for several binary modulation schemes can be calculated as [40, Eq. (12)]

$$\bar{P}_b = \frac{q^p}{2\Gamma(p)} \int_0^\infty \gamma^{p-1} \exp(-q\gamma) F(\gamma) d\gamma \tag{2.25}$$

In (2.25), four possible combinations of constants p and q have the capability to represent several forms of binary modulation schemes [40]. Importantly, the modulation techniques popular for the outdoor FSO transmission are binary phase shift keying (BPSK), differential binary phase shift keying (DBPSK), and binary frequency shift keying (BFSK).

2.3.2.1 Exact Analysis

The end-to-end CDF $F(\gamma)$ of the dual hop scheme (obtained by putting $\gamma_{th} = \gamma$ in outage probability expression (2.22)) is substituted in (2.25) to obtain the closed-form expression for the BER as follows:

$$\begin{aligned}\bar{P}_b = & \frac{q^p}{2\Gamma(p)} \int_0^\infty \gamma^{p-1} \exp(-q\gamma) d\gamma - \frac{q^p \chi_1}{2\Gamma(p)} \frac{(g)^{\frac{1}{2}} (h)^{-\frac{(\xi^2+\alpha\mu-\alpha)}{2r} - \frac{1}{2}}}{(2\pi)^{\frac{g+h-2}{2}}} \\ & \times \left(\frac{c}{\bar{\gamma}_1}\right)^{\frac{\xi^2+\alpha\mu-\alpha}{2r}} \int_0^\infty \gamma^{\frac{\xi^2+\alpha\mu-\alpha}{2r}+p-1} \exp\left[-\frac{\gamma}{\bar{\gamma}_1}\right] \mathrm{G}_{0,g+h}^{g+h,0}\left[B_2 \gamma^h \middle| \begin{matrix} - \\ \mathrm{d}_1 \end{matrix}\right] d\gamma\end{aligned} \tag{2.26}$$

where $B_2 = \left(\frac{\mathrm{B}}{g}\right)^g \left(\frac{c}{\bar{\gamma}_1 h}\right)^h$. Plugging $\int_0^\infty \gamma^{p-1} \exp(-q\gamma) d\gamma = q^{-p}\Gamma(p)$ and using [35, Eq. (07.34.21.0088.01)] to solve the integral involved in (2.26), we have

$$\bar{P}_b = \frac{1}{2} - \chi_1 \frac{(g)^{\frac{1}{2}} (h)^{(-\frac{\xi^2+\alpha\mu-\alpha}{2r} - 1 + s')}}{(2\pi)^{\frac{g+2h-3}{2}}} \left(\frac{c}{\bar{\gamma}_1}\right)^{\frac{(\xi^2+\alpha\mu-\alpha)}{2r}} \frac{1}{\left(q + \frac{1}{\bar{\gamma}_1}\right)^{s'}} \mathrm{G}_{0,g+h}^{g+h,h}\left[B_3 \middle| \begin{matrix} \mathrm{d}_2 \\ \mathrm{d}_1 \end{matrix}\right] \tag{2.27}$$

In (2.27), the constant $s' = \frac{(\xi^2+\alpha\mu-\alpha)}{2r} + p$, whereas the parameter d_2 consists of h terms in the series $[\frac{1-s'}{h}, .., \frac{h-s'}{h}]$. The argument, on the other hand, is given by $B_3 = \left(\mathrm{B}/g\right)^g \left(c/\bar{\gamma}_1 \left(q + \frac{1}{\bar{\gamma}_1}\right)\right)^h$.

2.3.2.2 Asymptotic Analysis

Similar to outage probability, the BER expression can be further simplified for the case when the average SNR becomes arbitrarily small. Using the expansion of Meijer-G function, the asymptotic approximate expression for the average BER at high SNR values can be obtained as

$$\bar{P}_b \underset{\bar{\mu}_r \gg 1}{\approx} \frac{1}{2} - \chi_1 \frac{(g)^{\frac{1}{2}}(h)^{(-\frac{\xi^2+\alpha\mu-\alpha}{2r}-1+s')}}{(2\pi)^{\frac{g+2h-3}{2}}} \left(\frac{c}{\bar{\gamma}_1}\right)^{\frac{\xi^2+\alpha\mu-\alpha}{2r}}$$
$$\times \frac{1}{\left(q+\frac{1}{\bar{\gamma}_1}\right)^{s'}} \sum_{j=1}^{g+h} \left[\left(\frac{g}{\mathrm{B}}\right)^{g}\left(\frac{\bar{\gamma}_1}{c}\left(q+\frac{1}{\bar{\gamma}_1}\right)\right)^{h}\right]^{-\mathrm{d}_{1,j}}$$
$$\times \prod_{i=1}^{g+h} \Gamma(\mathrm{d}_{1,i} - \mathrm{d}_{1,j}) \prod_{i=1}^{h} \Gamma(1 + \mathrm{d}_{1,j} - \mathrm{d}_{2,i}) \tag{2.28}$$

2.3.3 *Ergodic Capacity Analysis*

Ergodic capacity plays a crucial role in evaluating the performance of wireless communication systems, including FSO communication. It is a fundamental measure that quantifies the average data rate a system can achieve over a fading channel in the long term, given that coding schemes are used for removing the error caused in the channel. This metric provides valuable insights into the reliability and efficiency of the system. In the context of FSO communication, where atmospheric turbulence and dynamic environmental conditions can affect the optical channel, ergodic capacity becomes particularly important.

The significance of ergodic capacity lies in its ability to capture the statistical behavior of the channel over extended periods, offering a comprehensive understanding of the system's performance in real-world scenarios. By assessing the ergodic capacity of an FSO communication link, researchers and engineers can make well-informed decisions regarding modulation schemes, coding techniques, and system parameters. These decisions aim to optimize data transmission rates while ensuring reliable communication. In wireless systems, including FSO, where channel conditions can vary unpredictably, a deep understanding of ergodic capacity is crucial for designing robust and adaptive communication strategies. Incorporating ergodic capacity considerations into the design process allows for the development of communication systems that not only meet performance requirements under diverse channel conditions but also enhance spectral efficiency. This makes them highly suitable for meeting the ever-evolving demands of wireless communication.

As per the Shannon-Hartley law, the bit rate R with zero error can be transmitted in bandwidth of B Hz using various coding schemes. The threshold SNR γ_{th} is defined as $\gamma_{th} = e^{2R} - 1$. The ergodic capacity C ($C < R$) of the communication system is the maximum achievable data rate of a wireless fading channel given error-free transmission. Ergodic capacity is defined as $\bar{C} \triangleq \mathbb{E}[\log_2(1+\delta\gamma)]$ as presented in [41]

$$\bar{C} = \int_0^\infty \frac{1}{\ln(2)}(1+\delta\gamma)^{-1} F^c(\gamma) d\gamma \tag{2.29}$$

The authors of [42] have derived the value of constant δ for optical systems. In the case of IM/DD schemes, the value of δ can be determined by using a capacity bound, which is $\delta = \frac{e}{2\pi}$, due to the average-to-peak-power constraint. To enhance error performance compared to IM/DD techniques, coherent optical demodulation schemes are utilized on FSO links. These coherent demodulation schemes can be classified as homodyne or heterodyne detection schemes based on their carrier phase sensitivity and the presence of beat frequency. Both homodyne and heterodyne techniques extract information from the frequency and phase of the incoming optical wave, eliminating the need for an average-to-peak-power constraint. Based on this principle, $\delta = 1$ for coherent demodulation schemes is presented in [42], which indicates the constant involved in the capacity expression. Additionally, $F^c(\gamma)$ represents the complementary cumulative distribution function (CCDF) of the overall system.

2.3.3.1 Exact Analysis

The Meijer-G counterparts of $(1+\delta\gamma)^{-1} = \mathrm{G}_{1,1}^{1,1}\left[\delta\gamma \middle| \begin{smallmatrix} 0 \\ 0 \end{smallmatrix}\right]$ and $\exp\left(-\frac{\gamma}{\bar{\gamma}_1}\right) = \mathrm{G}_{0,1}^{1,0}\left[\frac{\gamma}{\bar{\gamma}_1} \middle| \begin{smallmatrix} - \\ 0 \end{smallmatrix}\right]$ with the help of [35, Eq. (07.34.03.0271.01) and Eq. (07.34.03.0228.01)], respectively, can be plugged into (2.29). In addition to this, the CCDF $F^c(\gamma) = 1 - F(\gamma)$ can be obtained from (2.22). This gives a form of integration of multiplication of three Meijer-G functions which can be evaluated using [43, Eq. (2.3)], utilizing which, the closed-form expression for the ergodic capacity can be obtained as

$$\begin{aligned}\bar{C} = \;& \frac{1}{\ln(2)}\chi_1 \bar{\gamma}_1 (g)^{\frac{1}{2}} (c)^{\frac{\xi^2+\alpha\mu-\alpha}{2r}} \\ & \times \frac{(h)^{-\frac{(\xi^2+\alpha\mu-\alpha)}{2}-\frac{1}{2}}}{(2\pi)^{\frac{g+h-2}{2}}} \mathrm{H}_{1,0:1,1:0,g+h}^{0,1:1,1:g+h,0}\left[\delta\bar{\gamma}_1, \left(\frac{\mathrm{B}}{g}\right)^g \left(\frac{c}{h}\right)^h \middle| \begin{matrix} (-\frac{a_1}{2r}; 1, h)\ (0,1) & - \\ - & (0,1)\ (\mathrm{d}_1, 1) \end{matrix}\right]\end{aligned} \tag{2.30}$$

where $\mathrm{H}^{0,m_1:m_2,n_2:m_3,n_3}_{p_1,q_1:p_2,q_2:p_3,q_3}\left[\mathcal{A}_1, \mathcal{A}_2\Big|\Big|\Big|\right]$ denotes the bivariate Fox's-H function defined in [44, Eq. (2.57)] and $a_1 = \xi^2 + \alpha\mu - \alpha$. Utilizing the definition of Mellin-Barnes type of contour integral, the bivariate Fox's-H function can be evaluated using the MATLAB software package as suggested in [45]. Another MATLAB implementation has been provided in [46], which is designed for bivariate Meijer-G function but can be adjusted to solve for the bivariate Fox's H function.

2.3.3.2 Asymptotic Analysis

The expression for the ergodic capacity must be recognized in a similar manner as the outage probability and BER. To accomplish this, the CCDF $F^c(\gamma)$ can be approximated for small arguments by using the expansion of the Meijer-G function, as described in [36].

$$F^c(\gamma) \underset{\bar{\mu}_r \gg 1}{\approx} \frac{(g)^{\frac{1}{2}}(h)^{-\frac{(\xi^2+\alpha\mu-\alpha)}{2r}-\frac{\chi_1}{2}}}{(2\pi)^{\frac{g+h-2}{2}}} \exp\left[-\frac{\gamma}{\bar{\gamma}_1}\right]\left(\frac{c\gamma}{\bar{\gamma}_1}\right)^{\frac{\xi^2+\alpha\mu-\alpha}{2r}}$$
$$\times \sum_{j=1}^{g+h}\left[\left(\frac{g}{\mathrm{B}}\right)^g\left(\frac{\bar{\gamma}_1 h}{c\gamma}\right)^h\right]^{-d_{1,j}} \prod_{i=1}^{g+h} \Gamma(d_{1,i} - d_{1,j}) \tag{2.31}$$

The CCDF in (2.31) can be substituted in (2.29) to further express the ergodic capacity as

$$\bar{C} \underset{\bar{\mu}_r \gg 1}{\approx} \frac{\chi_1}{\ln(2)} \frac{(g)^{\frac{1}{2}}(h)^{-\frac{(\xi^2+\alpha\mu-\alpha)}{2r}-\frac{1}{2}}}{(2\pi)^{\frac{g+h-2}{2}}}\left(\frac{c}{\bar{\gamma}_1}\right)^{\frac{\xi^2+\alpha\mu-\alpha}{2r}}$$
$$\times \sum_{j=1}^{g+h}\left[\left(\frac{g}{\mathrm{B}}\right)^g\left(\frac{\bar{\gamma}_1 h}{c}\right)^h\right]^{-d_{1,j}} \prod_{i=1}^{g+h} \Gamma(d_{1,i} - d_{1,j})$$
$$\times \int_0^\infty \frac{\gamma^{(hd_{1,j}+\frac{a_1}{2r})}}{(1+\delta)} \exp\left(-\frac{\gamma}{\bar{\gamma}_1}\right) d\gamma \tag{2.32}$$

With the help of [31, Eq. (383.10)], a closed-form solution to the integral in (2.32) can be obtained, which provides with the asymptotic representation of ergodic capacity as

$$\bar{C} \underset{\bar{\mu}_r \gg 1}{\approx} \frac{\chi_1}{\ln(2)} \exp\left(\frac{1}{\delta\bar{\gamma}_1}\right) \sum_{j=1}^{g+h}\left(\frac{1}{\delta}\right)^{hd_{1,j}+\frac{a_1}{2r}+1} B_3^{-d_{1,j}} \Gamma\left(hd_{1,j}+1+\frac{a_1}{2r}\right)$$

$$\times \prod_{i=1}^{g+h} \Gamma(d_{1,i} - d_{1,j}) \Gamma\Big((-hd_{1,j} - \frac{a_1}{2r}), \quad \frac{1}{\delta \bar{\gamma}_1}\Big) \tag{2.33}$$

where $\Gamma(w, r)$ represents the upper incomplete Gamma function [31, Eq. (8.350.2)] and the constant B_3 is obtained as $B_3 = (\frac{g}{\mathrm{B}})^g (\frac{\bar{\gamma}_1 h}{c})^h$.

2.4 Performance Analysis of Mixed RF/FSO DF Relaying System Model

When the relay node acts as DF node, it receives the incoming signal, decodes it, and then again encodes it for transmission. The equivalent SNR for the DF system can be expressed as

$$\gamma_{DF} = \min(\gamma_1, \gamma_2) \tag{2.34}$$

The outage probability can be derived for the DF relay operation by using the following relationship:

$$P_{out,DF} = F_{\gamma_1}(\gamma_{th}) + F_{\gamma_2}(\gamma_{th}) - F_{\gamma_1}(\gamma_{th}) F_{\gamma_2}(\gamma_{th}) \tag{2.35}$$

The required CDF for the RF link can be easily computed as follows:

$$F_{\gamma_1}(x) = \int_0^x f_{\gamma_1}(\gamma_1) d\gamma_1 \tag{2.36}$$

Solving the above integral for the PDF given in (2.4), we have

$$F_{\gamma_1}(x) = 1 - \exp\Big[-\frac{x}{\bar{\gamma}_1}\Big] \tag{2.37}$$

Similarly, the CDF on the FSO link undergoing $\alpha - \mu$ fading can be calculated as

$$F_{\gamma_2}(x) = \int_0^x f_{\gamma_2}(\gamma_2) d\gamma_2 \tag{2.38}$$

Placing the PDF from (2.17) into (2.38), we have

$$F_{\gamma_2}(x) = \frac{\xi^2}{r\Gamma(\mu)} \left(\frac{\mu}{\Omega_r A_o \hat{I}_a}\right)^{\frac{\xi^2+\alpha\mu-\alpha}{2}} W_{\frac{\alpha\mu-\xi^2-\alpha}{2\alpha}, \frac{\xi^2-\alpha\mu}{2\alpha}} \left(\frac{\mu}{\hat{I}_a^{\alpha}}\right) \int_0^x y^{\frac{\xi^2+\alpha\mu-\alpha}{2r}-1} \exp \\ \times \Big[-\mathrm{B} y^{\frac{\alpha}{r}}\Big] dy \tag{2.39}$$

As an intermediate step, the above integral can be expressed assuming $y^{\frac{\alpha}{r}}=z$ as

$$F_{\gamma_2}(x)=\frac{\xi^2}{r\Gamma(\mu)}\left(\frac{\mu}{\Omega_r A_o \hat{I}_a}\right)^{\frac{\xi^2+\alpha\mu-\alpha}{2}} W_{\frac{\alpha\mu-\xi^2-\alpha}{2\alpha},\frac{\xi^2-\alpha\mu}{2\alpha}}\left(\frac{\mu}{\hat{I}_a^{\alpha}}\right)\int_0^{x^{\frac{\alpha}{r}}} z^{\frac{\xi^2+\alpha\mu-3\alpha}{2\alpha}} \exp\Big[-\mathrm{B}z\Big]dz \tag{2.40}$$

A closed-form solution to the above integral can be evaluated using [31, Eq. (3.381.1)]

$$F_{\gamma_2}(x)=\frac{\xi^2}{r\Gamma(\mu)}\left(\frac{\mu}{\Omega_r A_o \hat{I}_a}\right)^{\frac{\xi^2+\alpha\mu-\alpha}{2}} W_{\frac{\alpha\mu-\xi^2-\alpha}{2\alpha},\frac{\xi^2-\alpha\mu}{2\alpha}}\left(\frac{\mu}{\hat{I}_a^{\alpha}}\right)\mathrm{B}^{-(\frac{\xi^2+\alpha\mu-3\alpha}{2\alpha}+1)} \times\gamma\left(\frac{\xi^2+\alpha\mu-3\alpha}{2\alpha}+1,\mathrm{B}x^{\frac{\alpha}{r}}\right) \tag{2.41}$$

The results from (2.37) and (2.41) can be substituted into (2.35) to derive the closed-form expression for the outage probability for the DF mixed RF/FSO relaying systems under $\alpha-\mu$ fading model as follows:

$$\begin{aligned}P_{out,DF}=&\frac{\xi^2}{r\Gamma(\mu)}\left(\frac{\mu}{\Omega_r A_o \hat{I}_a}\right)^{\frac{\xi^2+\alpha\mu-\alpha}{2}} W_{\frac{\alpha\mu-\xi^2-\alpha}{2\alpha},\frac{\xi^2-\alpha\mu}{2\alpha}}\left(\frac{\mu}{\hat{I}_a^{\alpha}}\right)\mathrm{B}^{-(\frac{\xi^2+\alpha\mu-3\alpha}{2\alpha}+1)}\\ &\times\gamma\left(\frac{\xi^2+\alpha\mu-3\alpha}{2\alpha}+1,\mathrm{B}\gamma_{th}^{\frac{\alpha}{r}}\right)\\ &+\left(1-\exp\Big[-\frac{\gamma_{th}}{\bar{\gamma}_1}\Big]\right)-\frac{\xi^2}{r\Gamma(\mu)}\left(\frac{\mu}{\Omega_r A_o \hat{I}_a}\right)^{\frac{\xi^2+\alpha\mu-\alpha}{2}}\\ &\times W_{\frac{\alpha\mu-\xi^2-\alpha}{2\alpha},\frac{\xi^2-\alpha\mu}{2\alpha}}\left(\frac{\mu}{\hat{I}_a^{\alpha}}\right)\mathrm{B}^{-(\frac{\xi^2+\alpha\mu-3\alpha}{2\alpha}+1)}\\ &\times\gamma\left(\frac{\xi^2+\alpha\mu-3\alpha}{2\alpha}+1,\mathrm{B}\gamma_{th}^{\frac{\alpha}{r}}\right)\left(1-\exp\Big[-\frac{\gamma_{th}}{\bar{\gamma}_1}\Big]\right)\end{aligned} \tag{2.42}$$

The outage probability expression for the DF relay takes into account for the fading on the RF link, turbulence on the optical link and the misalignment error.

2.5 Results and Discussions

The outage probability performance of the AF relay system is shown in Fig. 2.3 where various combinations of model parameters α and μ have been considered

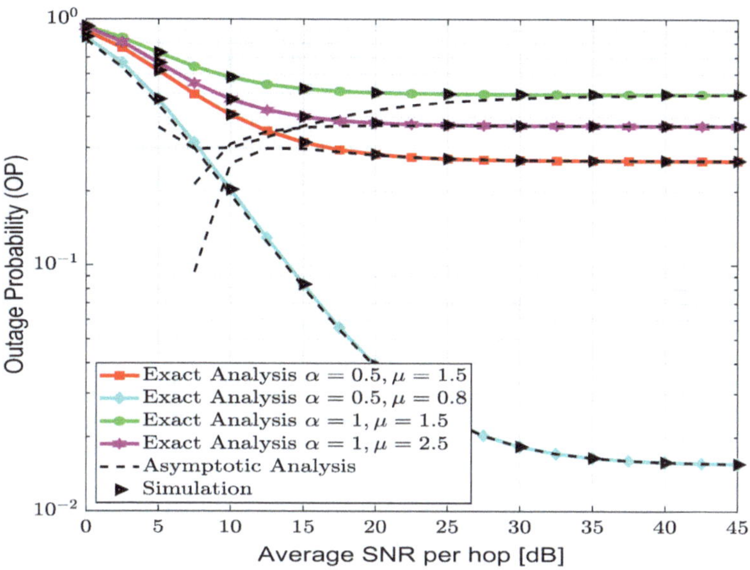

Fig. 2.3 Analysis of outage probability versus average SNR per hop for the proposed mixed RF/FSO dual hop relaying scheme

for representing the atmospheric turbulence. The Meijer-G function has been evaluated in MATLAB software. However, the Meijer-G function evaluation can be a complex step, and simplification provided in asymptotic analysis can be used to find the outage values. The derived expression for the outage probability in terms of elementary functions has also been plotted on the same axis, and the convergence between the exact and the asymptotic results is apparent in the plot. The results are also corroborated using the Monte Carlo simulation. The combinations of the parameters selected for the analysis are as follows $[(\alpha, \mu)] = [(0.5, 1.5), (0.5, 0.8), (1, 1.5), (1, 2.5)]$. Due to the dependency of the refraction structure parameter on the parameters, the best performance is obtained when $(\alpha, \mu) = (0.5, 0.8)$, which corresponds to weak turbulence among all the combinations of the parameters.

The BER analysis for the considered system model has been provided in Fig. 2.4. Here, the SNR on the RF link has been fixed to 10 dB. The electrical SNR on the FSO link has been varied to plot the characteristic curve. The basis of the analysis is the impact of pointing error. The parameter σ_s/a has been varied to increase the effect of pointing error for the mixed RF/FSO AF relaying system. Another outcome of the presented analysis is the dependency of the error rate on the kind of the demodulation scheme employed. To this end, the parameter r in the derived results has been varied, and $r = 2$ demonstrates IM/DD which gives worse results for each value of σ_s/a when compared to its counterpart $r = 1$ that corresponds to the coherent detection scheme. The complexity of the coherent form of heterodyne demodulation scheme comes with added advantage of enhanced

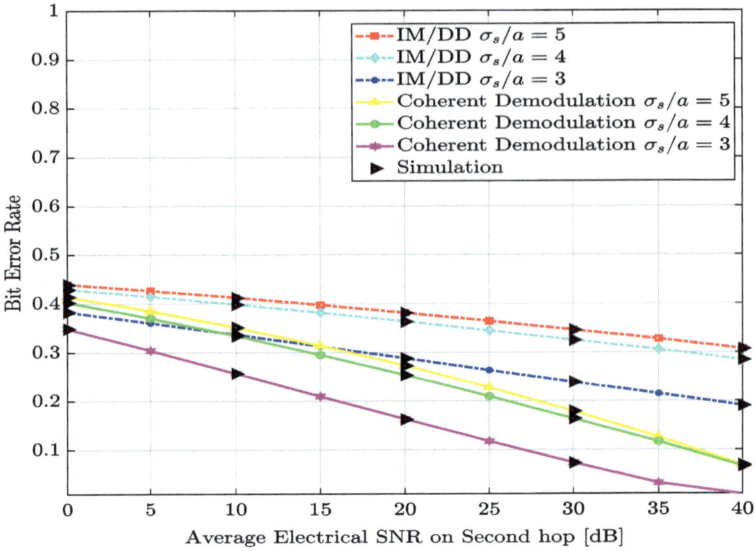

Fig. 2.4 Analysis of BER versus average electrical SNR on FSO link for varying values of pointing error jitter variance

coverage and reliability. The BER curve demonstrates this clearly. It can be deduced from the plot that even in worst form of pointing error, the coherent demodulation scheme performs better than the intensity based demodulation.

The capacity of the proposed system has been illustrated in Fig. 2.5. The capacity relates to the maximum achievable rate that can be obtained from the system. It is safe to assume that by adding FSO link as cooperation with the robust RF system, the data rate of the system can be enhanced. Since this chapter introduces the $\alpha - \mu$ model for representing various forms of atmospheric turbulence, the figure demonstrates the performance of the optical link using various combinations of the model parameters. The average SNR per hop has been increased to record the ergodic capacity values. The model parameter μ has been varied as $\mu = [0.5, 0.75, 1, 1, 25]$. For the given set of parameters, the changing values of the μ parameter suggest lower values of atmospheric turbulence, which results into higher achievable data rate due to large available bandwidth on the optical wireless channel.

In Fig. 2.6, the $\alpha - \mu$ modeling of the optical link has been demonstrated. Specifically, the values of model parameters α and μ have been evaluated by solving the Eqs. (2.13) and (2.14). This has been done by illustrating the fact that the $\alpha - \mu$ fading model can represent large range of turbulence. As it is well-known fact that the K-distribution is used to model severe turbulence conditions, the moments of the K-distribution have been utilized to evaluate the values of parameters α and μ, and outage probability of the system has been depicted for both forms of optical demodulation schemes. Similarly, the values of α and μ parameters for Gamma-Gamma distribution have been used to simulate moderate turbulence environment

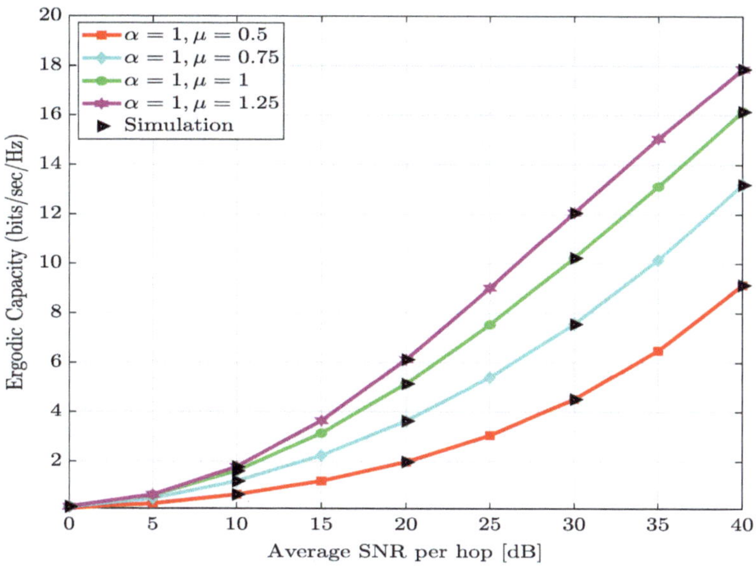

Fig. 2.5 Analysis of ergodic capacity versus SNR per hop for varying values of FSO link parameter μ

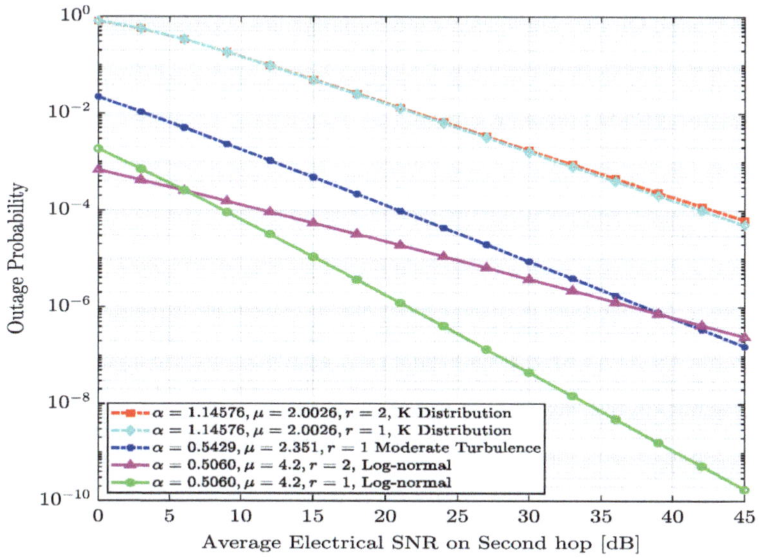

Fig. 2.6 Utilization of generalization capability of $\alpha - \mu$ turbulence model for generating severe to low fading optical channel

for the optical link. Also, weak turbulence is suggested using the log-normal fading model. Solving the moment based estimators for the α and μ parameters, it has been possible to find $\alpha = 0.5060$ and $\mu = 4.2$ against the log-normal turbulence model. The outage probability has been demonstrated for weak turbulence conditions for $r = 2$ and $r = 1$. It is evident from the plot that IM/DD performs inferior to the coherent form of optical demodulation.

2.6 Conclusion

This chapter presented the modeling of the optical channel using the $\alpha - \mu$ model. Since the model has been earlier only used to infer the fading on the RF links, a new mathematical description has been covered extensively in the chapter. Specifically, the moment based generator for the α and μ parameters has been used to cover a wider range of turbulence-induced fading on the FSO link. The mathematical model has been extended to derive a new PDF for the turbulence channel. The derived PDF has been unified corresponding to the kinds of the optical demodulation schemes that can be used in FSO systems. As an application to the derived theoretical results, study of mixed RF/FSO AF and DF relaying systems has been performed where the optical link undergoes optical fading over $\alpha - \mu$ model. Closed-form expressions have been derived for the outage probability, BER, and the ergodic capacity of the AF model. On the other hand, expression for the outage probability of the DF relaying system has also been presented. The derived results have been numerically evaluated to study the impact of turbulence and pointing error on the performance of the mixed RF/FSO relaying systems.

References

1. E. Lee, J. Park, D. Han, G. Yoon, Performance analysis of the asymmetric dual-hop relay transmission with mixed RF/FSO links. IEEE Photon. Technol. Lett. **23**(21), 1642–1644 (2011)
2. E. Zedini, I.S. Ansari, M.S. Alouini, Performance analysis of mixed nakagami-m and gamma-gamma dual-hop FSO transmission systems. IEEE Photonics J. **7**(1), 1–20 (2015)
3. H.G. Sandalidis, T.A. Tsiftsis, G.K. Karagiannidis, M. Uysal, BER performance of FSO links over strong atmospheric turbulence channels with pointing errors. IEEE Commun. Lett. **12**, 44–46 (2008)
4. Y. Tang, M. Brandt-Pearce, S.G. Wilson, Adaptive coding and modulation for hybrid FSO/RF systems, in *Proceedings of Forty-Third Asilomar Conference on Signals, Systems and Computers, Pacific Grove, CA* (2009), pp. 1644–1649
5. L. Yang, X. Gao, M.S. Alouini, Performance analysis of free-space optical communication systems with multiuser diversity over atmospheric turbulence channels. IEEE Photonics J. **6**(2), 1–17 (2014)
6. I.S. Ansari, F. Yilmaz, M.S. Alouini, Impact of pointing errors on the performance of mixed RF/FSO dual-hop transmission systems. IEEE Wireless Commun. Lett. **2**(3), 351–354 (2013)

7. I.S. Ansari, F. Yilmaz, M.S. Alouini, Performance analysis of free-space optical links over $\mathcal{M}$'$a$laga ($\mathcal{M}$) turbulence channels with pointing errors. IEEE Trans. Wirel. Commun. **15**(1), 91–102 (2016)
8. P. Wang, R. Wang, L. Guo, T. Cao, Y. Yang, On the performances of relay-aided FSO system over $\mathcal{M}$ distribution with pointing errors in presence of various weather conditions. Opt. Commun. **367**, 59–67 (2016)
9. T.A. Tsiftsis, H.G. Sandalidis, G.K. Karagiannidis, N.C. Sagias, Multihop free-space optical communications over strong turbulence channels. IEEE Int. Conf. Commun. **6**, 2755–2759 (2006)
10. X. Tang, Z. Wang, Z. Xu, Z. Ghassemlooy, Multihop free-space optical communications over turbulence channels with pointing errors using heterodyne detection. J. Lightwave Technol. **32**(15), 2597–2604 (2014)
11. E. Zedini, M.S. Alouini, On the performance of multihop heterodyne FSO systems with pointing errors. IEEE Photonics J. **7**(2), 1–10 (2015)
12. L.C. Andrews, R.L. Phillips, Laser Beam Propagation Through Random Media, 2nd edn. (SPIE Press, Bellingham, WA, 2005)
13. R.L. Phillips, L.C. Andrews, Measured statistics of laser light scattering in atmospheric turbulence. J. Opt. Soc. Am. **71**(12), 1440–144 (1981)
14. E. Jakeman, On the statistics of K-Distributed noise. J. Phys. Am. **13**, 31–48 (1980)
15. R. Barrios, F. Dios, Exponentiated weibull model for the irradiance probability density function of a laser beam propagating through atmospheric turbulence. Opt. Laser Technol. **45**, 13–20 (2013)
16. M.A. Kashani, M. Uysal, M. Kavehrad, A novel statistical channel model for turbulence-induced fading in free-space optical systems. IEEE/OSA J. Lightwave Technol. **33**(11), 2303–2312 (2015)
17. A.J. Navas, J.M.G. Balsells, J.F. Paris, A.P. Notario, A unifying statistical model for atmospheric optical scintillation, in *Numerical Simulations of Physical and Engineering Processes, Ch. 8* (2011)
18. M.D. Yacoub, The $\alpha - \mu$ distribution: a physical fading model for the stacy distribution. IEEE Trans. Veh. Technol. **56**(1), 27–34 (2007)
19. U.S. Dias, M.D. Yacoub, On the $\alpha - \mu$ autocorrelation and power spectrum functions: field trials and validation, in *GLOBECOM 2009—2009 IEEE Global Telecommunications Conference* (2009), pp. 1–6
20. K.P. Peppas, A simple, accurate approximation to the sum of gamma-gamma variates and applications in MIMO free-space optical systems. IEEE Photon. Technol. Lett. **23**(13), 839–841 (2011)
21. K.P. Peppas, F. Lazarakis, A. Alexandridis, K. Dangakis, Simple, accurate formula for the average bit error probability of multiple-input multiple-output free-space optical links over negative exponential turbulence channels. Opt. Lett. **37**(12), 3243–3245 (2012)
22. R. Boluda-Ruiz, A. García-Zambrana, B. Castillo-Vázquez, C. Castillo-Vázquez, Ergodic capacity analysis of decode-and-forward relay-assisted FSO systems over $\alpha - \mu$ fading channels considering pointing errors. IEEE Photonics J. **8**(1), 1–11 (2016)
23. Z. Ghassemlooy, W. Popoola, S. Rajbhandari, *Optical wireless communications: system and channel modeling with MATLAB*, 1st edn. (CRC Press, Boca Raton, 2012)
24. M.I. Petkovic, G.T. Dordevic, D.N. Milic, BER performance of IM/DD FSO system with OOK using APD receiver. Radioengineering **23**(1), 480–487 (2014)
25. C. Abou-Rjeily, Performance analysis of selective relaying in cooperative free-space optical systems. J. Light. Technol. **31**(18), 2965–2973 (2014). https://doi.org/10.1109/JLT.2013.2277191
26. T. Song, P. Y. Kam, A robust GLRT receiver with implicit channel estimation and automatic threshold adjustment for the free space optical channel with IM/DD. J. Light. Technol. **32**(4), 369–383 (2014). https://doi.org/10.1109/JLT.2013.2293137
27. S.M. Aghajanzadeh, M. Uysal, Multi-hop coherent free-space optical communications over atmospheric turbulence channels. IEEE Trans. Commun. **59**(6), 1657–1663 (2011)

28. X. Li, C. Chen, W.D. Zhong, D. Wu, MDPSK-based Non-equalization OFDM for coherent free-space optical communication. IEEE Photon. Technol. Lett. **26**(16), 1617–1620 (2014)
29. M.K. Simon, M.S. Alouini, *Digital Communication over Fading Channels*, 2nd edn. (Wiley, New York, 2005)
30. A.A. Farid, S. Hranilovic, Outage capacity optimization for free-space optical links with pointing errors. J. Lightwave Technol. **25**, 1702–1710 (2007)
31. I.S. Gradshteyn, I.M. Ryzhik, Table of Integrals, Series and Products (Academic Press, New York, 2000)
32. M.A. Al-Habash, L.C. Andrews, R.L. Phillips, Mathematical model for the irradiance PDF of a laser beam propagating through turbulent media. Opt. Eng. **40**(8), 1554–1562 (2001)
33. K. Kiasaleh, Performance of coherent DPSK free-space optical communication systems in K-Distributed turbulence. IEEE Trans. Commun. **54**(4), 604–607 (2006)
34. A. Upadhya, V.K. Dwivedi, G. Singh, Relay-aided free-space optical communications using $\alpha - \mu$ distribution over atmospheric turbulence channels with misalignment errors. Opt. Commun. **416**, 117–124 (2018)
35. I. Wolfram, Wolfram, Research, Mathematica Edition: Version 10.0. Champaign. (Wolfram Research, Inc., New York, 2010)
36. A.M. Mathai, R.K. Saxena, *Generalized Hypergeometric Functions with Applications in Statistics and Physical Sciences.* (Springer, Berlin, 1973), p. 348
37. M.S. Alouini, M.K. Simon, Dual diversity over correlated lognormal fading channels. IEEE Trans. Commun. **50**(12), 1946–1959 (2002)
38. X. Yi, M. Yao, X. Wang, MIMO FSO communication using sub-carrier intensity modulation over double generalized gamma fading. Opt. Commun. **382**, 64–72 (2017)
39. M.R. McKay, A. Zanella, I.B. Collings, M. Chiani, Error probability and SINR analysis of optimum combining in rician fading. IEEE Trans. Commun. **57**, 676–687 (2009)
40. I.S. Ansari, S. Al-Ahmadi, F. Yilmaz, M.-S. Alouini, H. Yanikomeroglu, A new formula for the BER of binary modulations with dual-branch selection over Generalized-K composite fading channels. IEEE Trans. Commun. **59**, 2654–2658 (2011)
41. A. Annamalai, R.C. Palat, J. Matyjas, Estimating ergodic capacity of cooperative analog relaying under different adaptive source transmission techniques, in *Proceedings of IEEE Sarnoff Symposium* (2010), pp. 1–5
42. A. Lapidoth, S.M. Moser, M.A. Wigger, On the capacity of free space optical intensity channels. IEEE Trans. Inf. Theory **55**, 4449–4461 (2009)
43. P. Mittal, K. Gupta, An integral involving generalized function of two variables. Proc. Indian Acad. Sci. Section A **75**(3), 117–123 (1972)
44. A. Kilbas, *H-Transforms: Theory and Applications. Analytical Methods and Special Functions* (Taylor and Francis, London, 2004)
45. K. Peppas, A new formula for the average bit error probability of dual hop amplify-and-forward relaying systems over generalized shadowed fading channels. IEEE Wireless Commun. Lett. **1**, 85–88 (2012)
46. H. Chergui, M. Benjillali, S. Saoudi, Performance analysis of project-and-forward relaying in mixed MIMO-pinhole and rayleigh dual-hop channel. IEEE Wireless Commun. Lett. **20**, 610–613 (2016)

Chapter 3
Interference Limited Dual Hop RF/FSO AF Relaying Over $\alpha - \mu$ Turbulence Channels

3.1 Introduction

The previous chapter established a framework for modeling the FSO channel using this $\alpha - \mu$ model. The considered mixed RF/FSO relaying system can be extended to consider the impact of interference on the RF link as well. The present chapter focuses on this. An interference limited cooperative relay system is analyzed. The results presented are more practical and general compared to when interference is neglected. For this, the optical wireless channel has been modeled using the $\alpha - \mu$ turbulence model, whereas the RF link has been assumed to undergo Rayleigh fading. The interference on the RF link, along with the pointing error and kind of optical demodulation, has been included in the analysis. The theoretical analysis in the chapter presents the analysis for the outage probability, BER, and ergodic capacity for the considered system.

3.1.1 Background

FSO technology can provide high data rates several Gbps by utilizing a large amount of license bandwidth. This enables the communication experts to support a greater number of users while consuming power [1]. For these, FSO systems have been considered as a strong candidate for backbone networks and last mile access networks, complementing RF wireless communication systems [1]. However, the transmission of optical signals in free space faces a challenging obstacle known as atmospheric turbulence or scintillation. This phenomenon arises due to rapid fluctuations in the phase and intensity of the optical wave caused by changes in the refractive index of the medium [2]. Additionally, another issue called pointing error affects the quality of service of FSO systems. It is caused by misalignment

A. Upadhya et al., *Free-Space Optical Communication Systems for Next Generation Networks*, https://doi.org/10.1007/978-3-031-74805-9_3

between the transmitter and receiver apertures, resulting from building sway due to dynamic wind loads or weak earthquakes [1].

3.1.2 Mixed RF/FSO Systems

A cooperative relaying technique has been proposed for FSO systems to overcome the problems faced during the practical implementation [3–8]. Cooperative relaying involves transmitting the information signal from the source terminal to the destination node through several intermediate terminals. This method utilizes the existing hardware of the intermediate mobile terminals to transmit the data. As a result, it can enhance the system's robustness to fading conditions, similar to diversity schemes, without increasing the physical size of the system. Cooperative relaying strategies can be classified as either AF or DF relay systems, depending on their complexity and operation. In addition, mixed RF/FSO cooperative systems have become popular as they offer a practical solution to the limited coverage issue in wireless communication systems. The mixed RF/FSO relaying scheme is an asymmetric scheme where both hops experience different levels of fading. Moreover, these systems can multiplex multiple RF users on the high bandwidth FSO channel, thereby increasing the system capacity. Reference [3] presents the impact of turbulence on the outage performance of mixed RF/FSO systems using SIM.

3.1.3 Effect of Interference on Wireless Communication

Although the effects of scintillation and pointing error on the FSO systems have been extensively addressed, analysis of the effect caused by interference on FSO systems has been inadequately treated. Furthermore, such investigation is important considering that FSO systems have to co-exist with different forms of communication systems. It is noteworthy that aggressive reuse of frequency bands for increasing spectral efficiency has resulted in interference for the wireless users from adjacent cells utilizing the same frequency components. However, extensive efforts have been made to investigate the impact of CCIs over wireless fading channels [9–21]. In research work [9], the authors have presented the end-to-end performance of multiuser AF relay networks assuming that the relay-to-destination link is effected by interference. Authors of [10] have performed the outage and error rate analysis of cellular systems in the presence of multiple CCIs. Recently, authors in [11] have presented a more realistic scenario for mixed RF/FSO asymmetric relaying system by including the effect of CCIs on the RF link. A DF user diversity mixed RF/FSO system has been analyzed in [12] to derive closed-form expressions for the outage probability and BER. On the other hand, a TWR AF mixed RF/FSO system with interference has been presented in [13]. A TWR DF scheme where user diversity is present and is governed by interference has been investigated in the research work [14]. The role of PRS for relay selection for overcoming the impact of interference has been presented by the authors of [15]. Here, the turbulence on the RF link has been modeled using the $\alpha - \mu$ irradiance model. A more generalized study has been presented in [16] where AF mixed RF/FSO relay system has been analyzed in the presence of CCIs where both RF and FSO links have been modeled using generalized distributions. The RF link has been assumed to undergo $\eta - \mu$ fading. Moreover, the nonzero boresight error component of the pointing error has been considered in the analysis. The impact of interference along with nonzero boresight error has been analyzed for TWR mixed RF/FSO relaying system in work [17]. Additionally, the impact of interference has been analyzed for non-orthogonal multiple access (NOMA) aided mixed RF/FSO relaying system in [18]. Interestingly, the joint impact of interference and hardware impairment has been presented in research work [19]. A similar work with double relay selection has been contributed in [20]. Significantly, the work in [21] considers deep learning for avoiding interference on the operation of mixed RF/FSO relaying systems.

3.2 Channel and System Model

An asymmetric mixed RF/FSO AF relaying system in the presence of N CCIs is considered for study as shown in Fig. 3.1. A practical scenario that relates with the present model can be providing high throughput signaling in an indoor environment on a FSO link. The RF tower assigns carriers in a dense urban environment that needs to be assigned to several users. The RF carrier can be sent toward a rooftop building from where it needs to be redirected toward the end user who wants to operate several devices, thus requiring a large amount of bandwidth. For this, a FSO link can be established from the rooftop, toward the window of the end user, where a destination node (router) can receive the optical signal. However, it is generally not possible for the rooftop relay node to function independently, and the presence of interference needs to be considered. On the first hop ($S - R$ RF link) of the relaying scheme, source node S communicates with the destination node D through the relay node R that is corrupted by N CCIs. At the relay node, the RF signal is converted into optical domain, which is then transmitted on the FSO link toward the destination node D on the $R - D$ link. Moreover, both the hops of the fixed gain relaying scheme are considered to be limited by additive white Gaussian noise (AWGN).

3.2.1 RF Link Statistics

The instantaneous signal-to-ratio noise (SNR) on the S-R link is assumed to follow Rayleigh distribution model [22]. The signal received at the relay node is

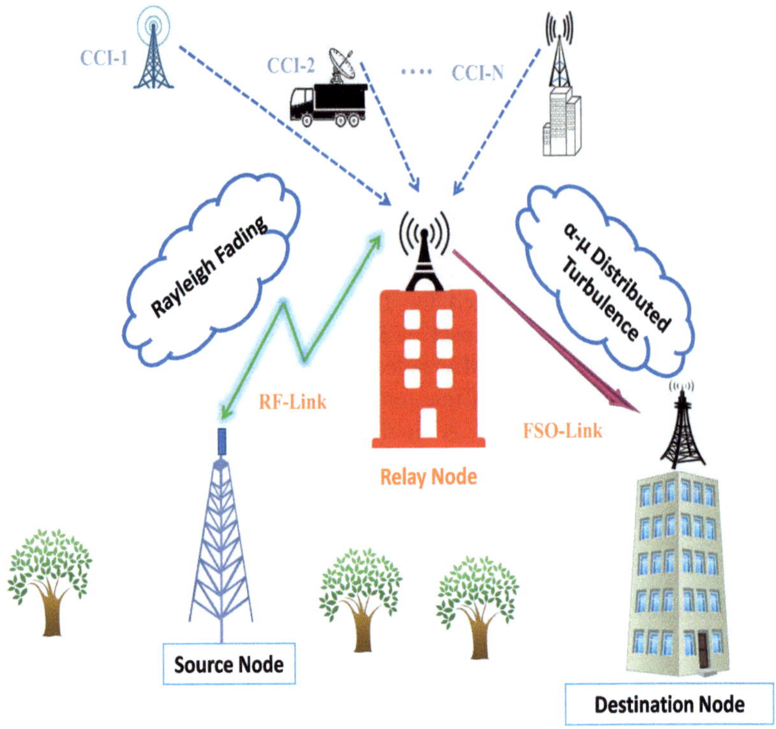

Fig. 3.1 System model for the mixed RF/FSO cooperative relaying model in the presence of interference

expressed as

$$r_{sr} = h_{sr}x + \sum_{i=1}^{N} h_i x_i + N_{01} \tag{3.1}$$

where x is the RF information signal, h_{sr} represents the Rayleigh distributed fading coefficient on the S-R link with power P_s, h_i denotes the fading amplitude of the ith CCI channel, and N_{01} is the AWGN with zero mean and variances σ_{sr}^2. The PDF of sum of random variables undergoing Rayleigh fading is given by Simon and Alouini [22]

$$f_{\gamma_1}(\gamma_1) = \frac{1}{\bar{\gamma}_1} \exp\left[-\frac{\gamma_1}{\bar{\gamma}_1} \right] \tag{3.2}$$

where $\bar{\gamma}_1$ is the average SNR on the S-R link. The PDF in (3.2) represents the statistics of signal on the RF link. The RF hop is limited by the presence of N identically distributed CCI with total interference-to-noise (INR) given by $\gamma_I = \sum_{i=1}^{N} \frac{{h_i}^2 P_I}{\sigma_{sr}^2}$. Considering each interferer following Rayleigh fading, the ordered statistics of the individual INRs is characterized by the chi-square distribution [10]

$$f_{\gamma_I}(\gamma_I) = \frac{\gamma_I^{N-1}}{\bar{\gamma}_I^N (N-1)!} \exp\left[-\frac{\gamma_I}{\bar{\gamma}_I}\right] \tag{3.3}$$

In (3.3), $\bar{\gamma}_I$ is the average INR per CCI link.

3.2.2 *FSO Link Statistics*

It is considered that after demodulation of the message on the RF link, the relay node converts the RF information into optical domain and after amplification transmits the same toward the destination node D where the received signal is converted back to electrical domain for demodulation. The forms of detection at node D considered can be either direct (i.e., IM/DD) or coherent demodulation scheme. The laser beam transmitting optical signal through the wireless channel is modeled over $\alpha - \mu$ distributed atmospheric turbulence with path loss and pointing error. The overall channel state on the FSO link is given as $I = I_a I_p I_l$. The factor I_l is the deterministic path loss, I_a is the turbulence-induced fading, and I_p denotes the pointing error caused by the misalignment between the transmitter and the receiver apertures. The atmospheric loss is determined by the Beers Lambert Law [23]

$$I_l(z) = \exp(-\sigma_1 z) \tag{3.4}$$

where $I_l(z)$ denotes the path loss over a propagation length of z, whereas σ_1 is the attenuation coefficient. Notably, generally, I_l is considered as a fixed scaling factor during the analysis over a long period of time, and no fluctuations are treated in its behavior. In the considered system model, turbulence-induced irradiance I_a on FSO link is modeled as $\alpha - \mu$ distribution, and the irradiance statistics is given by Peppas [24]

$$f_{I_a}(I_a) = \frac{\alpha\mu I_a^{\alpha\mu-1}}{\Gamma(\mu)\hat{\omega}_a^{\alpha\mu}} \exp\left[-\mu\left(\frac{I_a}{\hat{\omega}_a}\right)^{\alpha}\right] \tag{3.5}$$

where $\hat{\omega}_a = \sqrt[\alpha]{\mathbb{E}(I_a^{\alpha})}$ is the α-root mean value of atmospheric turbulence random variable, while α and μ are model parameters.

The fading statistics on the FSO link can be unified for both kinds of detection schemes as given below.

$$f(\gamma) = \frac{\xi^2}{r\Gamma(\mu)}\left(\frac{\mu}{\Omega_r A_o \hat{I}_a}\right)^{\frac{d}{2}} W_{\frac{\alpha\mu-\xi^2-\alpha}{2\alpha},\frac{\xi^2-\alpha\mu}{2\alpha}}\left(\frac{\mu}{\hat{I}_a^{\alpha}}\right)\gamma^{\frac{d}{2r}-1}\exp\left[-\mathrm{B}\gamma^{\frac{\alpha}{r}}\right] \tag{3.6}$$

where $r = 1$ corresponds to coherent detection technique, whereas $r = 2$ represents the IM/DD demodulation of optical signal with $\Omega_r = \frac{(\bar{\Psi}_r)^{\frac{1}{r}}}{\mathbb{E}[\mathrm{I}]}$. Notably, the preferred form of coherent demodulation is heterodyne detection.

3.2.3 *End-to-End SNR Statistics*

The destination signal in the presence of N CCIs at the relay node is given as [11]

$$x_d = (I\eta)^{\frac{r}{2}} G\left(h_{sr}x + \sum_{i=1}^{N} h_i x_i + N_{01} \right) + N_{02} \tag{3.7}$$

where each interferer operates with RF power, P_i denotes the power from $i = 1$th CCI, while the symbol transmitted is $\{x_i\}_{i=1}^{N}$. Also, N_{02} represents AWGN on the R-D link with zero mean and variance σ_{rd}^2. The optical-to-electrical conversion ratio at the destination node is denoted by η, while the overall irradiance is denoted by I and the constant G denotes fixed gain of the relay node. Assuming the presence of CSI at relay node R, the variable gain G provided by the AF scheme can be selected as [11]

$$G = \sqrt{\frac{P_t}{\mathbb{E}[h_{sr}^2]P_s + \sum_{i=1}^{N} \mathbb{E}[h_i^2]P_i + \sigma_{sr}^2}} \tag{3.8}$$

where P_t is the power transmitted from the relay node. Mathematical computation can be rephrased as

$$G^2 = \frac{\frac{P_t}{\sigma_{sr}^2}}{\frac{\mathbb{E}[h_{sr}^2]P_s}{\sigma_{sr}^2} + \frac{\sum_{i=1}^{N} \mathbb{E}[h_i^2]P_i}{\sigma_{sr}^2} + 1} \tag{3.9}$$

For the considered dual hop relay scheme in the presence of interferers, the overall signal-to-interference plus noise ratio (SINR), after assuming $C_1 = \bar{\gamma}_1 + \bar{\gamma}_I N + 1$, is given by Petkovic et al. [11]

$$\gamma = \frac{\gamma_1 \gamma_2}{\gamma_2 \gamma_I + \gamma_2 + C_1}. \tag{3.10}$$

3.3 Performance Analysis

A competent communication system is only feasible when all the important performance metrics are defined and approximately evaluated before actual implementation. For this purpose, it becomes indispensable to derive the analytical closed-form expressions for outage probability, BER, and ergodic capacity of the interference limited mixed RF/FSO AF relaying system. By utilizing the channel statistics defined in the previous section, various important performance metrics for the mixed RF/FSO system are derived to analyze system performance.

3.3.1 Outage Probability

Cooperative communication systems serve as the backbone of state-of-the-art connectivity and open the door for seamless transmission of data across vast distances. For such systems, outage probability emerges as an important metric, giving the likelihood of service degradation below acceptable levels. Understanding the importance of outage probability is of great interest in designing a reliable and robust wireless network that is able to meet the demands of current communication ecosystems. The outage probability can increase as a result of various impairments such as fading, interference, and noise. By estimating outage probability, network engineers can work out the system's ability to maintain congruous connectivity, sanctioning reliable data transmission crucial for applications ranging from voice calls to real-time multimedia streaming.

The evaluation of outage probability also makes the network planners well-equipped to allocate the resources optimally. The resources such as power, bandwidth, and spectrum are finite. Therefore, efficient allocation of such resources is essential for boosting the capacity and performance of wireless systems. Outage probability directly controls the resource allocation plan of action by delineating trade-offs between coverage, capacity, and reliability. By superintending resources based on outage probability performance, system planners can strike a balance between service availability and spectral efficiency, enhancing the performance of system while reducing operational costs. Here, the expression for outage probability has been derived for the considered system. The outage probability of the system provides the probability of the event when there is no connection between the transmitter and the receiver. If $P_r[x]$ denotes probability of event (x), then outage probability for the considered dual hop AF scheme effected by N interfering signals can be obtained as

$$
\begin{aligned}
P_{out} &= P_r\left[\frac{\gamma_1\gamma_2}{\gamma_2\gamma_I + \gamma_2 + C_1} < \gamma_{th}|\gamma_2\right] \\
&= \int_0^\infty \int_0^\infty Pr\left[\gamma_1 < \gamma_{th}\left(\gamma_I + 1 + \frac{C_1}{\gamma_2}\right)\right] f_{\gamma_2}(\gamma_2) f_{\gamma_I}(\gamma_I) d\gamma_2 d\gamma_I
\end{aligned}
\tag{3.11}
$$

To derive the closed-form expressions for the OP, the required CDF of RF hop can be expressed as

$$
F_{\gamma_1}(\gamma_1) = \int_0^\infty f_{\gamma_1}(\gamma_1) d\gamma_1 = 1 - \exp\left(\frac{-\gamma_1}{\bar{\gamma}_1}\right) \tag{3.12}
$$

Considering γ_2 and γ_I to be independent random variables, utilizing the definition $P_r[\gamma_1 < x] = F_{\gamma_1}(x)$, and putting the PDFs of γ_2 and γ_I in (3.11)

$$P_{out}=1-\frac{\xi^2}{r\Gamma(\mu)}\left(\frac{\mu}{\Omega_r A_o\hat{\omega}_a I_l}\right)^{\frac{d}{2}}W_{\frac{\alpha\mu-\xi^2-\alpha}{2\alpha},\frac{\xi^2-\alpha\mu}{2\alpha}}\left(\frac{\mu}{\hat{\omega}_a^{\alpha}}\right)\int_0^{\infty}\int_0^{\infty}\gamma_2^{\frac{d}{2r}-1}$$
$$\times\exp\left[-B\gamma_2^{\frac{\alpha}{r}}\right]\times\exp\left(\frac{-\gamma_{th}}{\bar{\gamma}_1}\right)\frac{\gamma_{\imath}^{N-1}}{\bar{\gamma}_{\imath}^{N}(N-1)!}\exp\left[-\frac{\gamma_{\imath}}{\bar{\gamma}_{\imath}}\right]d\gamma_2 d\gamma_{\imath} \quad (3.13)$$

Further, expressing $\exp\left(-\frac{C_1\gamma_{th}}{\gamma_2\bar{\gamma}_1}\right)=G_{1,0}^{0,1}\left[\frac{\gamma_2\bar{\gamma}_1}{C_1\gamma_{th}}\middle|\begin{matrix}1\\-\end{matrix}\right]$ with the help of [25, Eq. (07.34.03.0046.01)] and $\exp\left[-B\gamma_2^{\frac{\alpha}{r}}\right]=G_{0,1}^{1,0}\left[B\gamma_2^{\frac{\alpha}{r}}\middle|\begin{matrix}-\\0\end{matrix}\right]$ as per [26, Eq. (8.4.3.1)] in (3.13)

$$P_{out}=1-\frac{\xi^2}{r\Gamma(\mu)}\left(\frac{\mu}{\Omega_r A_o\hat{\omega}_a I_l}\right)^{\frac{d}{2}}W_{\frac{\alpha\mu-\xi^2-\alpha}{2\alpha},\frac{\xi^2-\alpha\mu}{2\alpha}}\left(\frac{\mu}{\hat{\omega}_a^{\alpha}}\right)\exp\left(\frac{-\gamma_{th}}{\bar{\gamma}_1}\right)$$
$$\times\int_0^{\infty}\gamma_2^{\frac{d}{2r}-1}G_{1,0}^{0,1}\left[\frac{\gamma_2\bar{\gamma}_1}{C_1\gamma_{th}}\middle|\begin{matrix}1\\-\end{matrix}\right]G_{0,1}^{1,0}\left[B\gamma_2^{\frac{\alpha}{r}}\middle|\begin{matrix}-\\0\end{matrix}\right]d\gamma_2\int_0^{\infty}\frac{\gamma_{\imath}^{N-1}}{\bar{\gamma}_{\imath}^{N}(N-1)!}$$
$$\times\exp\left[-\frac{\gamma_{\imath}}{\bar{\gamma}_{\imath}}\right]d\gamma_{\imath} \quad (3.14)$$

By utilizing standard identities provided in [26, Eq. (2.24.1.1)] and [27, Eq. (3.382.2)] along with some mathematical manipulations, the outage probability of an interference limited mixed RF/FSO system with fixed gain relaying in the presence of pointing errors over $\alpha-\mu$ atmospheric turbulence channels is derived to be

$$P_{out}=1-\frac{\xi^2}{r\Gamma(\mu)}\left(\frac{\mu}{\Omega_r A_o\hat{\omega}_a I_l}\right)^{\frac{d}{2}}W_{\frac{\alpha\mu-\xi^2-\alpha}{2\alpha},\frac{\xi^2-\alpha\mu}{2\alpha}}\left(\frac{\mu}{\hat{\omega}_a^{\alpha}}\right)\frac{(g)^{\frac{1}{2}}(h)^{-\left(\frac{d+r}{2r}\right)}}{(2\pi)^{\frac{h+g-2}{2}}}\left(\frac{C_1}{\bar{\gamma}_1}\right)^{\frac{d}{2r}}$$
$$\times\frac{1}{(\bar{\gamma}_{\imath})^N}\gamma_{th}^{\frac{d}{2r}}\times\exp\left[-\frac{\gamma_{th}}{\bar{\gamma}_1}\right]\left[\frac{1}{\bar{\gamma}_{\imath}}+\frac{\gamma_{th}}{\bar{\gamma}_1}\right]^{-N}G_{0,g+h}^{g+h,0}\left[\mathcal{K}_1\gamma_{th}^{h}\middle|\begin{matrix}-\\k_1\end{matrix}\right] \quad (3.15)$$

where the ratio $\frac{h}{g}=\frac{\alpha}{r}$ determines the order of Meijer-G, the argument of Meijer-G function is derived to be $\mathcal{K}_1=\left(\frac{B}{g}\right)^{g}\left(\frac{C_1}{\gamma_1 h}\right)^{h}$, and the parameters $k_1=[\Delta(g,0),\quad \Delta(h,\frac{-d}{2r})]$, where $\Delta(k,a)=[a/k,(a+1)/k,\ldots,(a+k-1)/k]$, consist of $g+h$ terms.

3.3.2 *Asymptotic Outage Probability Analysis*

The exact derived analytical expression for outage probability fails to provide any insights, before the same is implemented in a software package and Meijer-G evaluation. A better way is to form an approximate expression for the exact results that may help in gaining quick insights. Noting the fact that the argument of Meijer-

G is inversely proportional to the SNR (i.e., $\mathcal{K}_1 \propto 1/\bar{\Psi}_r$), as $\bar{\Psi}_r \rightarrow \infty$ in (3.15), the argument $\mathcal{K}_1 \rightarrow 0$. A very interesting property of the Meijer-G function is its ability to be expressed in terms of elementary functions when the argument is too small or too large. In this case, the argument becomes small as the average SNR increases. For a practical system, this approximation is almost every time valid. This fact can be utilized to derive the high SNR approximation of outage probability, and the argument of Meijer-G function can be inverted by invoking the identity [27, Eq. (9.31.2)]. For large arguments, with the aid of [25, Eq. (07.34.06.0018.01)], the Meijer-G function is represented in series form in terms of elementary functions only that can be given as

$$P_{out} \underset{\bar{\Psi}_r \gg 1}{\approx} 1 - \frac{\xi^2}{r\Gamma(\mu)} \left(\frac{\mu}{\Omega_r A_o \hat{\omega}_a I_l} \right)^{\frac{d}{2}} W_{\frac{\alpha\mu-\xi^2-\alpha}{2\alpha}, \frac{\xi^2-\alpha\mu}{2\alpha}} \left(\frac{\mu}{\hat{\omega}_a^{\alpha}} \right) \frac{(g)^{\frac{1}{2}}(h)^{-(\frac{d+r}{2r})}}{(2\pi)^{\frac{h+g-2}{2}}} \left(\frac{C_1}{\bar{\gamma}_1} \right)^{\frac{d}{2r}}$$
$$\times \frac{1}{(\bar{\gamma}_I)^N} \left[\frac{1}{\bar{\gamma}_I} + \frac{\gamma_{th}}{\bar{\gamma}_1} \right]^{-N}$$
$$\times \exp\left[-\frac{\gamma_{th}}{\bar{\gamma}_1} \right] \sum_{j=1}^{g+h} (\mathcal{K}_1)^{\mathrm{k}_{1,j}} \gamma_{th}^{h\mathrm{k}_{1,j}+\frac{d}{2r}} \prod_{i=1, i\neq j}^{g+h} \Gamma(\mathrm{k}_{1,i} - \mathrm{k}_{1,j}), \tag{3.16}$$

where $k_{1,j}$ denotes jth term in k_1 sequence presented in (3.16). Another important contribution of the approximate expression is the fact that it can be rephrased to provide coding gain and diversity order. For this, some more approximations can be applied to. Further, high SNR approximation $\bar{\gamma}_1 \rightarrow \infty$ can be applied in (3.15) to yield

$$\frac{\Gamma(N)}{(\bar{\gamma}_I)^N (N-1)!} \left[\frac{1}{\bar{\gamma}_I} + \frac{\gamma_{th}}{\bar{\gamma}_1} \right]^{-N} \underset{\bar{\gamma}_1 \rightarrow \infty}{\approx} \frac{\Gamma(N)}{(N-1)!} = 1 \tag{3.17}$$

The outage probability expression can be further simplified for the case when $\bar{\gamma}_1 \rightarrow \infty$ to express $\exp\left(\frac{-\gamma_1}{\bar{\gamma}_1}\right) \approx 1 - \left(\frac{\gamma_1}{\bar{\gamma}_1}\right)$ in (3.15). Moreover, assuming $x_{1,j} = hk_{1,j} + \frac{d}{2r}$ and $x_{2,j} = (g\alpha/r)k_{1,j}$, and by applying the aforementioned approximations, the OP can be expressed in terms of important parameters for the case of $\bar{\gamma}_1$, $\bar{\Psi}_r \rightarrow \infty$ and $\bar{\Psi}_r = \kappa\bar{\gamma}_1$ as

$$P_{out} \underset{\kappa\bar{\gamma}_1\bar{\gamma}_1, \bar{\Psi}_r \rightarrow \infty}{=} 1 - \mathcal{G}_1 \sum_{j=1}^{g+h} \left(\frac{\Lambda_j}{\kappa^{x_{2,j}}} \right) \gamma_{th}^{x_{1,j}} \tag{3.18}$$

where

$$\mathcal{G}_1 = \frac{\xi^2}{r\Gamma(\mu)} \left(\frac{\mu}{\Omega_r A_o \hat{\omega}_a I_l} \right)^{\frac{d}{2}} W_{\frac{\alpha\mu-\xi^2-\alpha}{2\alpha}, \frac{\xi^2-\alpha\mu}{2\alpha}} \left(\frac{\mu}{\hat{\omega}_a^{\alpha}} \right) \frac{(g)^{\frac{1}{2}}(h)^{-(\frac{d}{2r}+\frac{1}{2})}}{(2\pi)^{\frac{h+g-2}{2}}} C_1^{\frac{d}{2r}} \tag{3.19}$$

while the Λ_j involved in (3.18) is defined as

$$\Lambda_j = (\bar{\gamma}_1)^{-(x_{1,j}+x_{2,j})}\left[\frac{\mu}{2g}\left(\frac{x}{A_o\omega_a}\right)^{\alpha}\right]^{gk_{1,j}}\left(\frac{C_1}{h}\right)^{hk_{1,j}}\prod_{i=1,i\neq j}^{g+h}\Gamma(\mathrm{k}_{1,i}-\mathrm{k}_{1,j})$$

It is evident that the dominant factor influencing the asymptotic expression of outage probability is determined by the $\min(t_1, t_2)$, where t_1 represents the g-terms of k_1 and t_2 accounts for the remaining h-terms of parameter k_1.

3.3.3 *Coding Gain and Diversity Order Analysis*

In modern communication networks, achieving reliable and efficient data transmission over wireless channels is of greater importance. Two involved metrics in this trailing are diversity order and coding gain. These metrics play a significant role in determining the system's strength against channel impairments and noise, thereby improving its overall performance. Understanding and optimizing diversity order and coding gain have become imperative tasks in the design and analysis of wireless communication systems. Diversity order refers to the rate at which the outage probability decreases with increasing SNR in a communication system employing diversity techniques. In a cooperative relaying system, due to hopping strategy, the diversity order is generated without addition of additional hardware. It quantifies the system's ability to combat fading and multipath effects, which are prevalent in wireless channels. A higher diversity order signifies enhanced resilience to channel impairments, leading to improved reliability and coverage.

On the other hand, coding gain measures the improvement in SNR achieved by using error-correcting codes. It reflects the redundancy introduced by the coding scheme to mitigate errors introduced during transmission. By effectively combating channel noise and distortion, coding gain enhances the system's error performance and spectral efficiency. In this context, a comprehensive evaluation of diversity order and coding gain is requisite for optimizing system design parameters, such as modulation schemes, coding techniques, and diversity strategies. Moreover, their analysis provides valuable insights into the trade-offs between performance, complexity, and bandwidth utilization in wireless communication systems. This paper aims to explore the principles, significance, and optimization strategies associated with diversity order and coding gain in wireless communication systems, shedding light on their pivotal roles in ensuring reliable and efficient data transmission.

In the following, the system performance in terms of diversity order and coding gain has been presented using an analytical framework. The diversity order G_d is defined as the negative asymptotic slope of the outage curve versus SNR on a log scale [28]. On the other hand, coding gain G_c denotes the amount of shift in the SNR on the outage curve relative to reference curve of $(\bar{\gamma}_1)^{-G_d}$. The OP expression can be approximated in the high SNR regime given by the relation [29] $P_{out}\approx\left(G_c\bar{\gamma}_1\right)^{-G_d}$.

The diversity order G_d of the proposed mixed RF/FSO fixed gain relaying scheme in the presence of multiple interferers is given by

$$G_d = \frac{\alpha}{2r}\left(1 - \frac{d}{2r}\right) + \frac{d}{2r} + 1 \tag{3.20}$$

The coding gain G_c of the considered mixed RF/FSO fixed gain relaying scheme over $\alpha - \mu$ atmospheric turbulence channels, in the presence of multiple interferer, is derived to be

$$G_c = \left(\mathcal{G}_2 \gamma_{th}\right)^{\overline{\frac{\alpha}{2r}(1-\frac{d}{2r})+\frac{d}{2r}+1}^{\,-1}} \tag{3.21}$$

where

$$\mathcal{G}_2 = \frac{\xi^2}{r\Gamma(\mu)}\left(\frac{\mu}{\Omega_r A_o \hat{\omega}_a I_l}\right)^{\frac{d}{2}} W_{\frac{\alpha\mu-\xi^2-\alpha}{2\alpha},\frac{\xi^2-\alpha\mu}{2\alpha}}\left(\frac{\mu}{\hat{\omega}_a^{\alpha}}\right)\frac{C_1 r}{\alpha}\Gamma\left[\frac{r}{\alpha}\left(\frac{d}{2r} - 1\right)\right]$$
$$\times \left[\frac{\mu}{2}\left(\frac{1}{A_o\omega_a}\right)^{\alpha}\right]^{\frac{r}{2}(1-\frac{d}{2r})} \kappa^{\frac{\alpha}{2}(\frac{d}{2r}-1)-\frac{d}{2r}} \tag{3.22}$$

The diversity order G_d of the considered relay-aided transmission depends upon the value of fading parameters α, μ, type of demodulation rho, and the amount of pointing error. Interestingly, G_d does not depend upon the RF link parameters and is controlled only by the wireless optical link transmission characteristics. From the coding gain G_c developed in (3.21), it can be concluded that similar to G_d, coding also depends upon optical link of the dual hop AF relay scheme.

3.3.4 BER Analysis

BER of digital wireless communication channel is an important performance metric relating to the reliability of the network. BER holds predominant significance in wireless communication systems as it serves as a very major performance benchmark reflecting the quality of data transmission. BER gauges the probability of erroneous bit reception due to interference, noise, fading, and other channel impairments. In wireless networks where data integrity and reliability are predominant, achieving a low BER is essential to ensure accurate decoding at the receiver. A high BER can lead to degraded communication quality, reduced throughput, and ultimately, impaired user experience. Consequently, BER serves as a key criterion for assessing the effectiveness of various communication techniques, including modulation schemes, coding strategies, and diversity techniques. A digital wireless systems can achieve enhanced spectral efficiency, extended coverage, and improved overall performance by optimizing BER, making it a critical parameter in the design, analysis, and optimization of modern wireless communication systems. Authors in [30] provide a closed-loop formula for determining the average BER for a variety of binary modulation schemes

$$\bar{P}_b = \frac{q^p}{2\Gamma(p)} \int_0^\infty x^{p-1} \exp(-qx) F(x) dx \tag{3.23}$$

where various combinations of parameters p and q determine the modulation schemes involved and $F(\gamma)$ is the end-to-end channel CDF. Substituting $\gamma_{th} = \gamma$ in (3.15), the exact CDF can be easily obtained. Moreover, substituting the end-to-end CDF, thus obtained, for the mixed RF/FSO system into (3.23), expression for the average BER of the system can be further expressed as

$$\bar{P}_b = \frac{q^p}{2\Gamma(p)} \mathcal{Z}_1 - \frac{q^p \xi^2}{2r\Gamma(p)\Gamma(\mu)(N!)} \left(\frac{\mu}{\Omega_r A_o \hat{\omega}_a I_l}\right)^{\frac{d}{2}} \times W_{\frac{\alpha\mu-\xi^2-\alpha}{2\alpha}, \frac{\xi^2-\alpha\mu}{2\alpha}} \left(\frac{\mu}{\hat{\omega}_a^\alpha}\right) \frac{1}{\bar{\gamma}_I^N} \frac{(g)^{\frac{1}{2}}(h)^{-\left(\frac{d+r}{2r}\right)}}{(2\pi)^{\frac{h+g-2}{2}}} \quad \mathcal{Z}_2 \tag{3.24}$$

where $\mathcal{Z}_1 = \int_0^\infty \gamma^{p-1} \exp(-q\gamma) d\gamma = q^{-p}\Gamma(p)$, while the second integral is defined as

$$\mathcal{Z}_2 = \int_0^\infty x^{\frac{d}{2r}+p-1} \exp\left[-x\left(q + \frac{1}{\bar{\gamma}_1}\right)\right]\left[\frac{1}{\bar{\gamma}_I} + \frac{x}{\bar{\gamma}_1}\right]^{-N} \mathrm{G}_{0,g+h}^{g+h,0}\left[\mathcal{K}_1 x^h \middle| \overline{k_1}\right] dx \tag{3.25}$$

Representing $\left[\frac{1}{\bar{\gamma}_I} + \frac{\gamma_{th}}{\bar{\gamma}_1}\right]^{-N} = \frac{\bar{\gamma}_I^N}{N!} \mathrm{G}_{1,1}^{1,1}\left[\frac{\gamma\bar{\gamma}_I}{\bar{\gamma}_1} \middle| {1-N \atop 0}\right]$ and expressing $\exp\left[-\gamma\left(q + \frac{1}{\bar{\gamma}_1}\right)\right] = \mathrm{G}_{0,1}^{1,0}\left[\gamma\left(q+\frac{1}{\bar{\gamma}_1}\right)\middle| \overline{0}\right]$ with the aid of [25, Eq. (07.34.03.0271.01)] and [25, Eq. (07.34.03.0228.01)], respectively, the integral in (3.25) can be further expressed as

$$\mathcal{Z}_2 = \int_0^\infty x^{\frac{d}{2r}+p-1} \mathrm{G}_{0,1}^{1,0}\left[x\left(q+\frac{1}{\bar{\gamma}_1}\right)\middle| \overline{0}\right] \mathrm{G}_{1,1}^{1,1}\left[\frac{x\bar{\gamma}_I}{\bar{\gamma}_1}\middle| {1-N \atop 0}\right] \mathrm{G}_{0,g+h}^{g+h,0}\left[\mathcal{K}_1 x^h \middle| \overline{k_1}\right] dx$$

Furthermore, by representing each Meijer-G function in terms of Fox's H function on account of [25, Eq (07.34.26.0008.01)], $\mathcal{Z}_2$ can be solved by applying the identity [31, Eq. (2.3)]. If $z_1 = \frac{\bar{\gamma}_I}{\bar{\gamma}_1}\left(q + \frac{1}{\bar{\gamma}_1}\right)^{-1}$ and $z_2 = \mathcal{K}_1\left(q + \frac{1}{\bar{\gamma}_1}\right)^{-h}$ are the arguments of the bivariate Fox-H function, the closed-form expression for the BER of the considered system model can be written as

$$\bar{P}_b = \frac{1}{2} - \frac{q^p \xi^2}{2r\Gamma(p)\Gamma(\mu)} \left(\frac{\mu}{\Omega_r A_o \hat{\omega}_a I_l}\right)^{\frac{d}{2}} \frac{(g)^{\frac{1}{2}}}{(2\pi)^{\frac{h+g-2}{2}}}$$
$$\times W_{\frac{\alpha\mu-\xi^2-\alpha}{2\alpha}, \frac{\xi^2-\alpha\mu}{2\alpha}} \left(\frac{\mu}{\hat{\omega}_a^\alpha}\right) \left(q + \frac{1}{\bar{\gamma}_1}\right)^{-(p+\frac{d}{2r})} \left(\frac{C_1}{\bar{\gamma}_1}\right)^{\frac{d}{2r}}$$
$$\times H^{0,1:1,1:g+h,0}_{1,0:1,1:0,g+h} \left[\begin{array}{c|c|c} 1-p-\frac{d}{2r}; 1, h & (1-N,1) & \text{———} \\ \text{———} & (0,1) & (k_1, [1]_{\text{length}(k_1)}) \end{array} \middle| z_1, z_2 \right] \tag{3.26}$$

The exact analysis results of the average BER have been achieved in terms of the bivariate Fox's H function defined in [31, Eq. (2.1)]. Efficient MATHEMATICA software algorithm to evaluate bivariate Meijer-G function has been provided in [30]. This algorithm can be modified to develop a computationally efficient method to compute bivariate Fox's H function. Additionally, author in [32] has reported a MATLAB implementation for evaluating Fox's H function of two variables. Moreover, similar to outage probability, the high SNR approximation of BER can be established to gain better insights about the error performance for the proposed relaying strategy. Making note of the asymptotic representation of Meijer-G function, the high SNR approximation of the end-to-end CDF can be given as

$$F(\gamma) \underset{\bar{\Psi}_r \gg 1}{\approxeq} 1 - \frac{\xi^2}{r\Gamma(\mu)} \left(\frac{\mu}{\Omega_r A_o \hat{\omega}_a I_l}\right)^{\frac{d}{2}} \frac{(g)^{\frac{1}{2}} (h)^{-(\frac{d+r}{2r})}}{(2\pi)^{\frac{h+g-2}{2}}} \left(\frac{C_1}{\bar{\gamma}_1}\right)^{\frac{d}{2r}}$$
$$\times W_{\frac{\alpha\mu-\xi^2-\alpha}{2\alpha}, \frac{\xi^2-\alpha\mu}{2\alpha}} \left(\frac{\mu}{\hat{\omega}_a^\alpha}\right) \frac{1}{\bar{\gamma}_I{}^N}$$
$$\times \exp\left[-\frac{\gamma}{\bar{\gamma}_1}\right] \left[\frac{1}{\bar{\gamma}_I} + \frac{\gamma}{\bar{\gamma}_1}\right]^{-N} \sum_{j=1}^{g+h} \mathcal{K}_1^{\mathrm{k}_{1,j}} \prod_{i=1, i\neq j}^{g+h} \Gamma(\mathrm{k}_{1,i} - \mathrm{k}_{1,j}) \gamma^{h\mathrm{k}_{1,j} + \frac{d}{2r}} \tag{3.27}$$

Plugging the approximation of CDF into (3.23), it can be deduced that

$$\bar{P}_b \approxeq \frac{q^p}{2\Gamma(p)} \int_0^\infty x^{p-1} \exp(-qx) dx - \frac{q^p}{2\Gamma(p)} - \frac{\xi^2}{r\Gamma(\mu)} \left(\frac{\mu}{\Omega_r A_o \hat{\omega}_a I_l}\right)^{\frac{d}{2}}$$
$$\times \frac{(g)^{\frac{1}{2}} (h)^{-(\frac{d+r}{2r})}}{(2\pi)^{\frac{h+g-2}{2}}} \left(\frac{C_1}{\bar{\gamma}_1}\right)^{\frac{d}{2r}}$$
$$\times W_{\frac{\alpha\mu-\xi^2-\alpha}{2\alpha}, \frac{\xi^2-\alpha\mu}{2\alpha}} \left(\frac{\mu}{\hat{\omega}_a^\alpha}\right) \frac{1}{\bar{\gamma}_I{}^N} \sum_{j=1}^{g+h} \mathcal{K}_1^{\mathrm{k}_{1,j}} \prod_{i=1, i\neq j}^{g+h} \Gamma(\mathrm{k}_{1,i} - \mathrm{k}_{1,j}) \gamma^{h\mathrm{k}_{1,j} + \frac{d}{2r}}$$

$$\times \int_0^\infty x^{p-1} \exp(-qx) \exp\Big[-\frac{x}{\bar{\gamma}_1}\Big]\Big[\frac{1}{\bar{\gamma}_I} + \frac{x}{\bar{\gamma}_1}\Big]^{-N} dx \tag{3.28}$$

Invoking [27, Eq. (3.381.4)], and the asymptotic expression of the average BER can be derived as

$$\begin{aligned}\bar{P}_b \underset{\bar{\Psi}_r, \bar{\gamma}_1 \gg 1}{\approx} & \frac{1}{2} - \frac{q^p \xi^2}{2\Gamma(p) r \Gamma(\mu)} \Big(\frac{\mu}{\Omega_r A_o \hat{\omega}_a I_l}\Big)^{\frac{d}{2}} \frac{(g)^{\frac{1}{2}} (h)^{-(\frac{d+r}{2r})}}{(2\pi)^{\frac{h+g-2}{2}}} \Big(\frac{C_1}{\bar{\gamma}_1}\Big)^{\frac{d}{2r}} \\ & \times W_{\frac{\alpha\mu - \xi^2 - \alpha}{2\alpha}, \frac{\xi^2 - \alpha\mu}{2\alpha}} \Big(\frac{\mu}{\hat{\omega}_a^\alpha}\Big) \\ & \times \sum_{j=1}^{g+h} \mathcal{K}_1^{\mathrm{k}_{1,j}} \Gamma(v_j) \prod_{i=1, i \neq j}^{g+h} \Gamma(\mathrm{k}_{1,i} - \mathrm{k}_{1,j}) (1/a_j^v) \mathrm{U}(v_j, v_j + 1 - N, (1/\bar{\gamma}_I))\end{aligned} \tag{3.29}$$

It can be seen in the derived closed-form expression for the BER that there is now no requirement of accessing the bivariate Fox's H function. The BER asymptotic expression is indeed only in terms of elementary functions. Importantly, no important design component is missing in the expression. The asymptotic results also consider interference, RF fading, pointing error, turbulence, and path loss, while providing BER components for a practical range of SNR operations.

3.3.5 Ergodic Capacity Analysis

Another important performance indicator of the relay-assisted communication system is the capacity offered by the overall system when limited by noisy channel for reliable transmission. Ergodic capacity is defined as [33]

$$\bar{C} = \frac{1}{2\ln(2)} \int_0^\infty (1 + \delta\gamma)^{-1} F^c(\gamma) d\gamma \tag{3.30}$$

where $\delta = \frac{e}{2\pi}$ [34] for IM/DD reception technique and $\delta = 1$ for coherent detection, while $F^c(\gamma)$ denotes the end-to-end complementary CDF (CCDF) of the AF relay system. It is important to mention that (3.30) provides the lower bound on the ergodic capacity for IM/DD detection technique. By expressing $(1 + \delta\gamma)^{-1} = \mathrm{G}_{1,1}^{1,1}\left[\delta\gamma \middle| \begin{smallmatrix} 0 \\ 0 \end{smallmatrix}\right]$ with the help of [25, Eq. (07.34.03.0271.01)] and by plugging the requisite terms in (3.30), a closed-form solution of the capacity can be formulated as: Making use of this definition by putting $(1+\delta\gamma)^{-1} = \mathrm{G}_{1,1}^{1,1}\left[\delta\gamma \middle| \begin{smallmatrix} 0 \\ 0 \end{smallmatrix}\right]$ [25, Eq. (07.34.03.0271.01)] and CCDF into (21), we have

$$\bar{C} = \frac{1}{\ln(2)} \frac{\xi^2}{2\Gamma(\mu)} \left(\frac{\mu}{\Omega A_o \hat{I}_a I_l}\right)^{\frac{d}{2}} W_{\frac{\alpha\mu-\xi^2-\alpha}{2\alpha}, \frac{\xi^2-\alpha\mu}{2\alpha}}\left(\frac{\mu}{\hat{I}_a^\alpha}\right)$$
$$\times \sum_{k=0}^{\infty}(-1)^k \binom{N+k-1}{k} \gamma_l^{N+k} \quad \gamma_1^{-k}$$
$$\times \int_0^\infty \gamma^{k+\frac{d}{4}} G_{0,1}^{1,0}\left[\frac{\gamma}{\bar{\gamma}_1}\middle| \begin{matrix} - \\ 0 \end{matrix}\right] G_{1,1}^{1,1}\left[\delta\gamma \middle| \begin{matrix} 0 \\ 0 \end{matrix}\right] \quad G_{0,g+h}^{g+h,0}\left[B_1 \gamma^h \middle| \begin{matrix} - \\ k_1 \end{matrix}\right] d\gamma \tag{3.31}$$

In (22), $\exp\left(-\frac{\gamma}{\bar{\gamma}_1}\right)$ has been represented in terms of Meijer-G function. The integral with product of three Meijer-G functions can be solved by converting each function into Fox-H generalized function using [25, Eq (07.34.26.0008.01)], and then applying [31, Eq. (2.3)], we have

$$\bar{C} = \frac{1}{\ln(2)} \frac{\xi^2}{2\Gamma(\mu)} \left(\frac{\mu}{\Omega A_o \hat{I}_a I_l}\right)^{\frac{d}{2}} W_{\frac{\alpha\mu-\xi^2-\alpha}{2\alpha}, \frac{\xi^2-\alpha\mu}{2\alpha}}\left(\frac{\mu}{\hat{I}_a^\alpha}\right)$$
$$\times \sum_{k=0}^{\infty}(-1)^k \binom{N+k-1}{k} \gamma_l^{N+k} \quad \gamma_1^{-k}$$
$$\times H_{1,0:1,1:0,g+h}^{0,1:1,1:g+h,0}\left[\delta\bar{\gamma}_1, B_1\bar{\gamma}_1^h \middle| \begin{matrix} (-k-\frac{d}{4}; 1, h)\ (0,1) \quad - \\ - \quad (0,1)\ (k_1, 1) \end{matrix}\right] \tag{3.32}$$

Efficient MATHEMATICA implementation of bivariate Meijer-G function has been provided in [30], which can be utilized to develop a computationally efficient algorithm to compute bivariate Fox-H function defined in [31, Eq. (2.1)]. Another research work [35] presents MATLAB implementation for evaluation of Fox-H function in two variables. Next, in order to relate with the capacity results quickly, the asymptotic analysis for the same is addressed. For the ***high SNR*** approximation, the capacity definition in (3.30) can be utilized by substituting the asymptotic expression of CCDF given as

$$F^c(\gamma) \underset{\psi_2 \gg 1}{\approxeq} \frac{\xi^2}{2\Gamma(\mu)} \left(\frac{\mu}{\Omega A_o \hat{I}_a I_l}\right)^{\frac{d}{2}} W_{\frac{\alpha\mu-\xi^2-\alpha}{2\alpha}, \frac{\xi^2-\alpha\mu}{2\alpha}}\left(\frac{\mu}{\hat{I}_a^\alpha}\right) \frac{(g)^{\frac{1}{2}}(h)^{-\frac{d+2}{4}}}{(2\pi)^{\frac{h+g-2}{2}}} \left(\frac{C_1}{\bar{\gamma}_1}\right)^{\frac{d}{2}}$$
$$\times \frac{\Gamma(N)}{(\bar{\gamma}_l)^N N!}\left[\frac{1}{\bar{\gamma}_l} + \frac{\gamma}{\bar{\gamma}_1}\right]^{-N} \exp\left[-\frac{\gamma}{\bar{\gamma}_1}\right]$$
$$\times \sum_{j=1}^{g+h}\left[\left(\frac{g}{B}\right)^g \left(\frac{\bar{\gamma}_1 h}{C_1}\right)^h\right]^{-k_{1,j}} \gamma^{hk_{1,j}+\frac{d}{4}} \prod_{i=1}^{g+h} \Gamma(k_{1,i}-k_{1,j}) \tag{3.33}$$

Putting the asymptotic ***high SNR*** representation of the CCDF into (3.30), and making use of [Ref binomial exp neg power], we have

$$\bar{C} \underset{\psi_2 \gg 1}{\approx} \frac{\xi^2}{2\ln(2)\Gamma(\mu)} \left(\frac{\mu}{\Omega A_o \hat{I}_a I_l} \right)^{\frac{d}{2}} W_{\frac{\alpha\mu - \xi^2 - \alpha}{2\alpha}, \frac{\xi^2 - \alpha\mu}{2\alpha}} \left(\frac{\mu}{\hat{I}_a^\alpha} \right) \frac{(g)^{\frac{1}{2}} (h)^{-\frac{d+2}{4}}}{(2\pi)^{\frac{h+g-2}{2}}} \left(\frac{C_1}{\bar{\gamma}_1} \right)^{\frac{d}{2}}$$
$$\times \frac{\Gamma(N)}{N!} \sum_{m=0}^{\infty} \sum_{j=1}^{g+h} \left[\left(\frac{g}{\mathrm{B}} \right)^g \left(\frac{\bar{\gamma}_1 h}{C_1} \right)^h \right]^{-\mathrm{k}_{1,j}}$$
$$\times \prod_{i=1}^{g+h} \Gamma(\mathrm{k}_{1,i} - \mathrm{k}_{1,j}) (-1)^m a^m \binom{N+m-1}{m} \int_0^\infty \frac{\gamma^{h\mathrm{k}_{1,j} + \frac{d}{4} + m}}{1 + \delta\gamma} \left[1 + \gamma \frac{\bar{\gamma}_I}{\bar{\gamma}_1} \right]^{-N}$$
$$\times \exp\left[-\frac{\gamma}{\bar{\gamma}_1} \right] d\gamma \tag{3.34}$$

Utilizing [27, Eq. (3.383.10)] to solve the integral in (3.34), we have

$$\left(\begin{array}{l} \bar{C} \underset{\psi_2 \gg 1}{\approx} \frac{\xi^2}{2\ln(2)\Gamma(\mu)} \left(\frac{\mu}{\Omega A_o \hat{I}_a I_l} \right)^{\frac{d}{2}} \\ \qquad W_{\frac{\alpha\mu - \xi^2 - \alpha}{2\alpha}, \frac{\xi^2 - \alpha\mu}{2\alpha}} \left(\frac{\mu}{\hat{I}_a^\alpha} \right) \frac{(g)^{\frac{1}{2}} (h)^{-\frac{d+2}{4}}}{(2\pi)^{\frac{h+g-2}{2}}} \left(\frac{C_1}{\bar{\gamma}_1} \right)^{\frac{d}{2}} \frac{\Gamma(N)}{N!} \end{array} \right)$$
$$\times \sum_{m=0}^{\infty} \sum_{j=1}^{g+h} \left[\left(\frac{g}{\mathrm{B}} \right)^g \left(\frac{\bar{\gamma}_1 h}{C_1} \right)^h \right]^{-\mathrm{k}_{1,j}} \prod_{i=1}^{g+h} \Gamma(\mathrm{k}_{1,i} - \mathrm{k}_{1,j}) (-1)^m a^m \binom{N+m-1}{m}$$
$$\times (\delta)^{-(\frac{d}{4} + h\mathrm{k}_{1,j} + m + 1)} \times \exp\left[-\frac{1}{\delta\bar{\gamma}_1} \right] \Gamma(\upsilon) \Gamma\left(1 - \upsilon, \frac{1}{\delta\bar{\gamma}_1} \right) \tag{3.35}$$

where $a = \frac{\bar{\gamma}_I}{\bar{\gamma}_1}$ and $\upsilon = -(\frac{d}{4} + h\mathrm{k}_{1,j} + m + 1)$. Understanding the asymptotic expression for ergodic capacity holds significant importance in the analysis and design of communication systems, particularly in scenarios with high SNR. The ergodic capacity-derived expression represents the maximum achievable data rate over fading channels when the channel fades are averaged out over time but in simpler form. By examining the asymptotic behavior of ergodic capacity at high SNR, insights can be gained into the system's performance under ideal conditions. This analysis enables engineers to optimize system parameters such as modulation schemes, coding techniques, and transmit power allocations to maximize data throughput while maintaining reliability. Furthermore, the asymptotic expression for ergodic capacity provides valuable guidelines for accessing the scalability and efficiency of communication systems as SNR increases, thereby facilitating the design of future-proofed and robust wireless networks.

3.4 Results and Discussion

In this section, some numerical examples are presented corresponding to the developed analytical framework. The choice of model parameters is kept arbitrary to prove the validity of the obtained results. The numerical results obtained in the present work are validated by computer-based simulations in MATLAB software. In Fig. 3.2, the outage probability performance for different combinations of $\alpha - \mu$ parameter values used for modeling the turbulence on the FSO link has been presented. The considered combinations are $(\alpha, \mu) = [(0.5, 0.8), (0.67, 0.8), (1, 0.8)]$. Among the considered model parameter values, the combination of $(\alpha, \mu) = (1, 0.8)$ demonstrates the most severe turbulence conditions, as the outage probability is the highest. The INR of $N = 2$ interfering signals has been assumed to be $\bar{\gamma}_I = -10\,\text{dB}$ each. On the same scale, the asymptotic analysis presented in the chapter has been depicted. Clearly, the approximate expressions from the asymptotic analysis follow well with the exact results for the practical range of signal power. As expected, it is evident from the plot that as the average SNR (per hop) is increased, the outage probability decreases implying better reliability of communication system.

Figure 3.3 demonstrates the outage probability for $\alpha = 0.8$ and $\mu_2 = 1.4$ as a function of average SNR per hop for a different number of interferers. In this plot, the average INR for each interferer has been set to $\bar{\gamma}_I = 25\,\text{dB}$. The increase in the number of signal degrading interferer leads to a reduction in the reliability of the overall link as the outage probability increases. The Monte Carlo simulation is in agreement with the exact results as per theoretical analysis.

The plot considering $p = 0.5$ and $q = 1$, which corresponds to bipolar phase shift keying (BPSK) modulation scheme, has been depicted in Fig. 3.4. The turbulence has been modeled using $(\alpha, \mu) = (0.5, 0.8)$, $(0.67, 0.8)$, and $(1, 0.8)$. The BER offered by the overall communication system decreases as the average

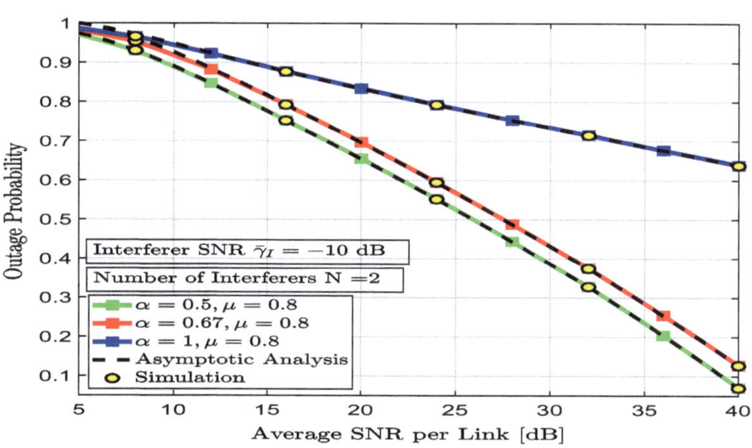

Fig. 3.2 Outage probability for different combinations of $\alpha - \mu$

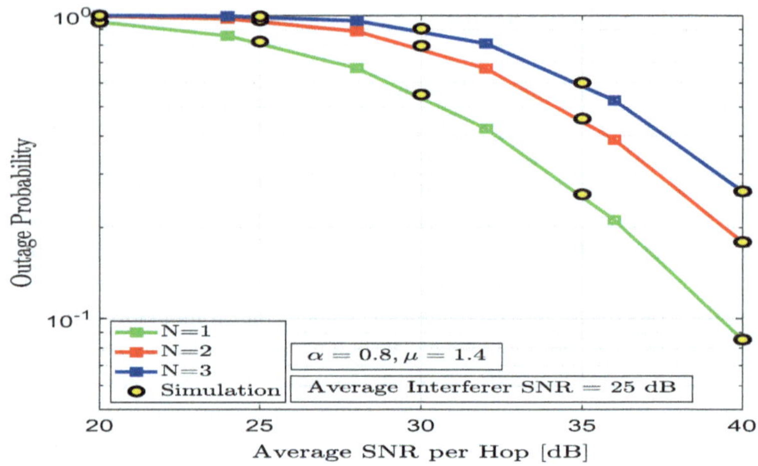

Fig. 3.3 Analysis of increase in the number of interferers N for the considered system model

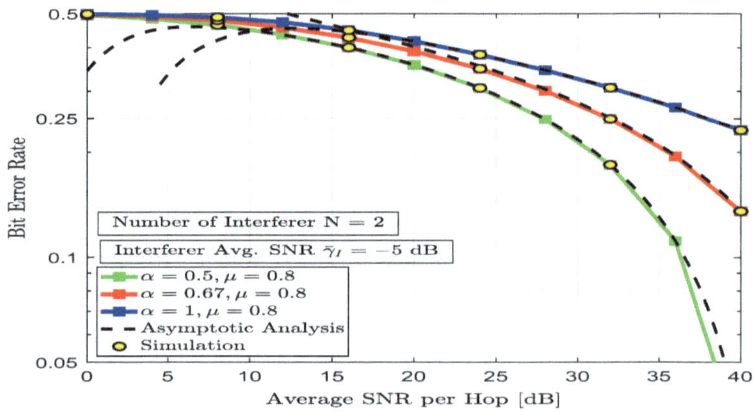

Fig. 3.4 BER analysis of the mixed RF/FSO cooperative relaying system for different sets of turbulence parameters α, μ

SNR per hop increases, as evident in the plot. It is therefore possible to enhance the power amplifier output to increase the power for reducing the BER, but only up to a given threshold. A range from 0 to 40 dB has been considered for calculating the BER for the mixed RF/FSO cooperative relaying system. To demonstrate the impact of interference, it has been considered that $N = 2$ CCIs are in the vicinity of the relay node. The average INR of each of these interferers has been assumed to be $\bar{\gamma}_I = -5$ dB. Combination of $\alpha, \mu = (1, 0.8)$ corresponds to the most severe turbulence condition on the optical link as the error rate is maximum for this set of parameters.

Similarly, the BER performance has been presented in Fig. 3.5 for a different number of interferers. As the number of interferer increases from $N = 1$ to $N = 3$,

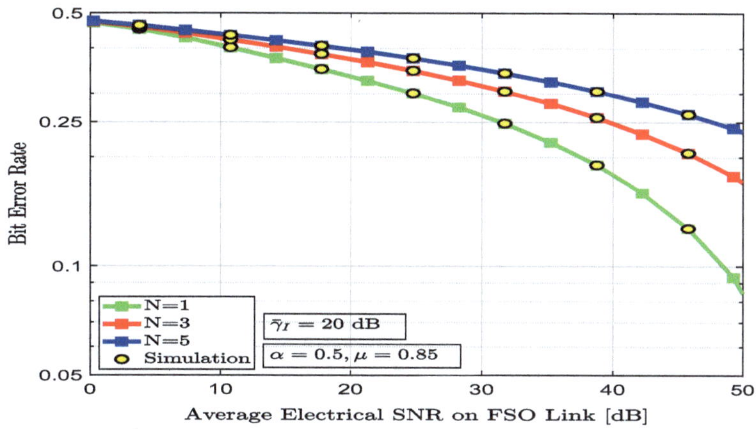

Fig. 3.5 BER performance for increase in the number of interferers at the relay node

the error rate offered by the system increases. The plot suggests what can happen when the number of operating devices increases nearby the relay node that is used to convert incoming RF signal into optical signal, after amplification. Therefore, the enhancement in error rate due to the availability of neighboring devices cannot be neglected. For fixed $\bar{\gamma}_I$ of 20 dB at $(\alpha, \mu) = (0.5, 0.85)$, it is evident in the plot that as the electrical SNR on the FSO link increases, the BER decreases. The results are verified using the Monte Carlo simulation. The main outcome of the plot is that in the analysis, the impact of interference cannot be neglected. This is because the reliability of the system is severely affected by interference along with atmospheric turbulence.

The throughput that can be harnessed from mixed RF/FSO system depends on the ergodic capacity of the network. For the considered practical system model, the ergodic capacity depends on the amount of fading on the radio link, optical power, alignment between the optical transmitter and receiver along with the interference that is present in the system. It is not possible for the system to operate in the absence of interference, and therefore quantification of the number and strength of interference becomes a key to the analysis. For demonstrating the performance of ergodic capacity, the electrical SNR on the FSO link is fixed to 33 dB in Fig. 3.6. The ergodic capacity is then plotted against the average SNR on the RF link. It is clear from the plot that as the average SNR increases, the capacity offered by the considered system increases. The turbulence on the FSO link has been modeled using $(\alpha, \mu) = [(0.5, 0.75), (0.67, 0.75), (1, 0.75)]$. The enhancement in the value of α parameter suggests from the plot that the irradiance fluctuations are more as the ergodic capacity for the case when $\alpha = 1$ is minimum in the plot. The interference INR has been assumed to be $\bar{\gamma}_I = -10$ dB. For validation of theoretical capacity results, the plot also demonstrates the Monte Carlo simulation results that are in agreement with the results. The asymptotic results that have been derived using

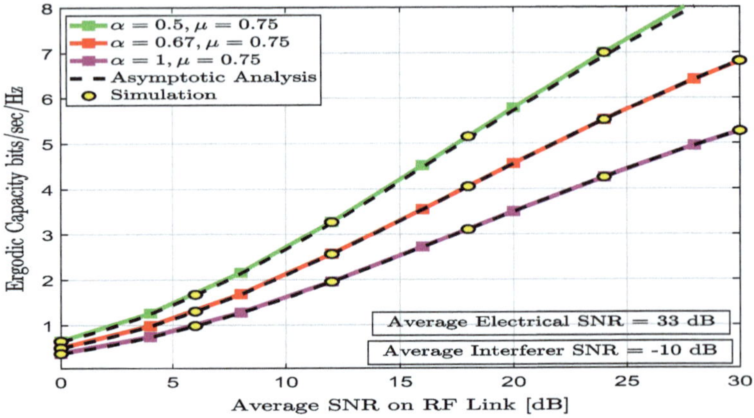

Fig. 3.6 Ergodic capacity of the mixed RF/FSO interference limited system for different combinations of model parameters

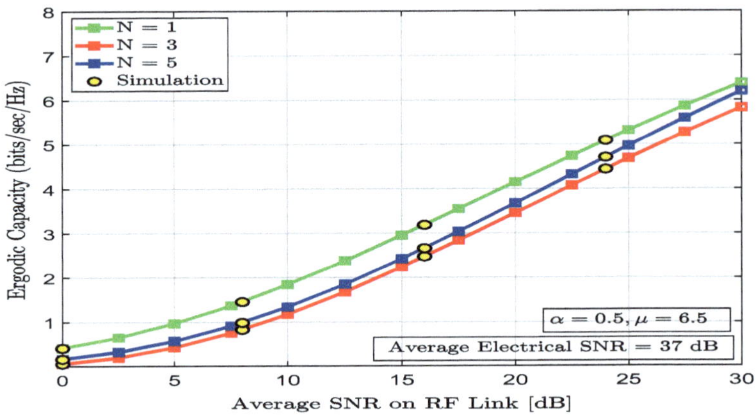

Fig. 3.7 Ergodic capacity performance of the considered system model for $N = [1, 2, 3]$

elementary functions have also been depicted in the figure. The asymptotic results meet with the exact results as the average SNR on the RF link increases.

As stated earlier, the impact of interference needs to be considered during the analysis of mixed RF/FSO systems. Figure 3.7 demonstrates the ergodic capacity of the network for different numbers of interferers $N = [1, 2, 3]$. As clearly depicted, the increase in the number of interferers leads to decrease in the capacity offered by the system. The theoretical results are confirmed using the Monte Carlo simulation.

On the other hand, if the ergodic capacity is plotted against the average electrical SNR on the FSO link, it is clear that the ergodic capacity increases with increase in electrical SNR in Fig. 3.8. However, for the purpose of analysis, signal power is varied on the first hop of communication. The average SNR on the RF link has been assumed to be $\bar{\gamma}_1 = [20, 30, 40]$. An increase in the average SNR on the RF link

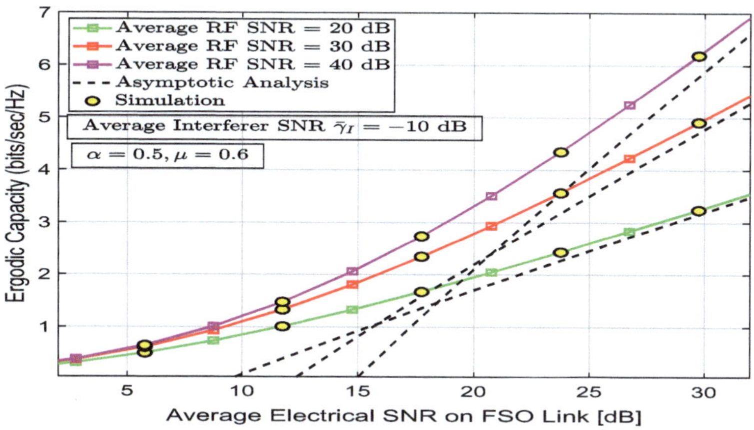

Fig. 3.8 Ergodic capacity analysis of the considered system model for different combinations of the signal power on the RF link

enhances the throughput of the system. This is an important fact that the system engineer needs to consider while designing a backhaul network. As the FSO link cannot cover long distances, the RF power can be tuned in such a manner so as to meet the throughput requirements in urban conditions. The results have been verified using Monte Carlo simulation, which are in agreement with exact theoretical results. The asymptotic analysis results also converge with the exact results as the electrical SNR on the optical link is increased.

3.5 Conclusion

This chapter explores the performance of RF/FSO AF relaying systems under interference limitations. Analytical expressions for outage probability, average BER, and ergodic capacity have been developed. These performance metrics have been rigorously examined in the high SNR regime. Furthermore, the asymptotic expression for outage probability is utilized to determine the coding gain and diversity order of the system as a whole. Numerical simulations are conducted to corroborate the mathematical framework proposed in this chapter, providing empirical validation of the presented model.

References

1. A.A. Farid, S. Hranilovic, Outage capacity optimization for free-space optical links with pointing errors. J. Lightwave Technol. **25**, 1702–1710 (2007)

2. Z. Ghassemlooy, W. Popoola, S. Rajbhandari, *Optical Wireless Communications: System and Channel Modeling with MATLAB* (CRC Press, New York, 2012)
3. E. Lee, J. Park, D. Han, G. Yoon, Performance analysis of the asymmetric dual-hop relay transmission with mixed RF/FSO links. IEEE Photon. Technol. Lett. **23**(21), 1642–1644 (2011)
4. E. Zedini, I.S. Ansari, M.S. Alouini, Performance analysis of mixed nakagami-m and gamma-gamma dual-hop FSO transmission systems. IEEE Photonics J. **7**(1), 1–20 (2015)
5. G. Djordjevic, M. Petkovic, A. Cvetkovic, G. Karagiannidis, Mixed RF/FSO relaying with outdated channel state information. IEEE J. Sel. Areas Commun. **33**(9), 1935–1948 (2015)
6. A.H. Abd El-Malek, A.M. Salhab, S.A. Zummo, M.S. Alouini, Security-reliability trade-off analysis for multiuser SIMO mixed RF/FSO relay networks with opportunistic user scheduling. IEEE Trans. Wirel. Commun. **15**(9), 5904–5918 (2016)
7. A. Touati, A. Abdaoui, F. Touati, M. Uysal, A. Bouallegue, On the effects of combined atmospheric fading and misalignment on the hybrid FSO/RF transmission. OSA J. Opt. Commun. Networking **8**(10), 715–725 (2016)
8. P. Wang, R. Wang, L. Guo, T. Cao, Y. Yang, On the performances of relay-aided FSO system over $\mathcal{M}$ distribution with pointing errors in presence of various weather conditions. Opt. Commun. **367**, 59–67 (2016)
9. A. Agustin, J. Vidal, Amplify-and-forward cooperation under interference-limited spatial reuse of the relay slot. IEEE Trans. Wirel. Commun. **7**(5), 1952–1962 (2008)
10. M.O. Hasna, M.S. Alouini, A. Bastami, E.S. Ebbini, Performance analysis of cellular mobile systems with successive co-channel interference cancellation. IEEE Trans. Wirel. Commun. **2**(1), 29–40 (2004)
11. M.I. Petkovic, A.M. Cvetkovic, G.T. Djordjevic, G.K. Karagiannidis, Outage performance of the mixed RF/FSO relaying channel in the presence of interference. Wireless Pers. Commun. **96**(2), 2999–3014 (2017)
12. A. Upadhya, V.K. Dwivedi, G. Singh, Multiuser diversity for mixed RF/FSO cooperative relaying in the presence of interference. Opt. Commun. **442**, 77–83 (2019)
13. A. Upadhya, V.K. Dwivedi, M.S. Alouini, Interference-limited mixed MUD-RF/FSO two-way cooperative networks over double generalized gamma turbulence channels. IEEE Commun. Lett. **442**, 1551–1555 (2019)
14. V.K. Tonk, A. Upadhya, P.K. Yadav, V.K. Dwivedi, Mixed mud-RF/FSO two way dcode and forward relaying networks in the presence of co-channel interference. Opt. Commun. **464**, 125415 (2020)
15. S.B. Bambiwal, A. Upadhya, R.S. Yaduvanshi, V.K. Pandey, Partial relay selection for combating the effects of co-channel interference in RF/FSO cooperative relaying. Opt. Commun. **475**, 126186 (2020)
16. A. Upadhya, V.K. Dwivedi, G.K. Karagiannidis, On the effect of interference and misalignment error in mixed RF/FSO systems over generalized fading channels. IEEE Trans. Commun. **68**(6), 3681–3695 (2020)
17. A. Upadhya, J. Gupta, V.K. Dwivedi, M.-S. Alouini, Impact of RF I/Q imbalance on interference-limited mixed RF/FSO TWR systems with non-zero boresight error. IEEE Wireless Commun. Lett. **10**(2), 416–420 (2021)
18. A. Upadhya, Investigation of mixed RF/FSO decode-and-forward NOMA cooperative relaying networks. Wireless Pers Commun. **124**, 2923–2938 (2022)
19. A. Goel, R. Bhatia, Joint impact of interference and hardware impairments on the performance of mixed RF/FSO cooperative relay networks. Opt. Quant Electron, Springer **53**(9), 530 (2021)
20. A. Goel, R. Bhatia, Double relay selection for combating the impact of interference and hardware impairments in mixed RF/FSO two way relay networks. Opt. Quant. Electron. **54**(11), 751 (2022)
21. A. Upadhya, GAN based channel estimation and interference cancellation for mixed RF/FSO cooperative relaying systems. Phys. Commun. **61**, 102199 (2023)
22. M.K. Simon, M.S. Alouini, *Digital Communication over Fading Channels*, 2nd edn. (Wiley, New York, 2005)

23. H.G. Sandalidis, T.A. Tsiftsis, G.K. Karagiannidis, M. Uysal, BER performance of FSO links over strong atmospheric turbulence channels with pointing errors. IEEE Commun. Lett. **12**, 44–46 (2008)
24. K.P. Peppas, A simple, accurate approximation to the sum of gamma-gamma variates and applications in MIMO free-space optical systems. IEEE Photon. Technol. Lett. **23**(13), 839–841 (2011)
25. I. Wolfram, *Wolfram, research, mathematica edition: Version 10.0. Champaign.* (Wolfram Research, Inc., New York, 2010)
26. O.I.M.A.P. Prudnikov, A. Brychkov, *Integrals and Series Special Functions* (CRC Press, New York, 1986)
27. I.S. Gradshteyn, I.M. Ryzhik, *Table of Integrals, Series and Products.* (Academic, New York, 2000)
28. M.A. Kashani, M. Safari, M. Uysal, Optimal relay placement and diversity analysis of relay-assisted free-space optical communication systems. IEEE/OSA J. Opt. Commun. Networking **5**(1), 32–47 (2013)
29. Z. Wang, G.B. Giannakis, A simple and general parameterization quantifying performance in fading channels. IEEE Trans. Commun. **51**, 1389–1398 (2003)
30. I.S. Ansari, S. Al-Ahmadi, F. Yilmaz, M.-S. Alouini, H. Yanikomeroglu, A new formula for the BER of binary modulations with dual-branch selection over Generalized-K composite fading channels. IEEE Trans. Commun. **59**, 2654–2658 (2011)
31. P. Mittal, K. Gupta, An integral involving generalized function of two variables. Proc. Indian Acad. Sci. Section A **75**(3), 117–123 (1972)
32. K.P. Peppas, A new formula for the average bit error probability of dual-hop amplify-and-forward relaying systems over generalized shadowed fading channels. IEEE Wireless Commun. Lett. **1**(2), 85–88 (2012)
33. A. Annamalai, R.C. Palat, J. Matyjas, Estimating ergodic capacity of cooperative analog relaying under different adaptive source transmission techniques. *Proceedings of IEEE Sarnoff Symposium* (2010), pp. 1–5
34. A. Lapidoth, S.M. Moser, M.A. Wigger, On the capacity of free space optical intensity channels. IEEE Trans. Inf. Theory **55**, 4449–4461 (2009)
35. H. Chergui, M. Benjillali, S. Saoudi, Performance analysis of project-and-forward relaying in mixed MIMO-pinhole and rayleigh dual-hop channel. IEEE Wireless Commun. Lett. **20**, 610–613 (2016)

Chapter 4
Reliability Analysis for Interference Limited Mixed MUD-RF/FSO Variable Gain AF Cooperative Relays Over D-GG Channel

4.1 Introduction

This chapter presents investigation of a multiuser hybrid system combining RF and FSO in an AF relay network, considering asymmetric channel fading conditions. The model accounts for CCIs affecting both relay and destination nodes. On the first hop of the variable gain AF strategy, multiple RF links facilitate information exchange between users and the relay node, while FSO links handle data transmission between the relay and destination nodes. The analysis assumes Nakagami-m fading for RF channels and D-GG fading for the wireless optical channel, incorporating the impact of pointing errors. Through this system model, exact expressions are derived for the outage probability and average BER. Additionally, the outage characteristics are explored under high SNR conditions to ascertain the coding gain and diversity order afforded by the system. These asymptotic findings inform a power optimization scheme, offering formulas for optimal power transmission at the transmitting nodes. Moreover, an approximate assessment of the ergodic capacity of the overall system is provided. Theoretical findings are substantiated through Monte Carlo simulations to validate their accuracy.

4.2 Background

The expansive unlicensed bandwidth available in FSO communication systems has garnered significant attention from researchers in recent years. FSO systems offer a range of advantages including rapid deployment, high security, and flexibility, which make them promising candidates for various communication applications. However, these benefits are constrained by factors such as atmospheric turbulence and pointing errors, which can introduce signal fluctuations and impair system performance. Atmospheric turbulence refers to the random variations in the refractive

A. Upadhya et al., *Free-Space Optical Communication Systems for Next Generation Networks*, https://doi.org/10.1007/978-3-031-74805-9_4

index of air, leading to scintillation effects that cause signal intensity fluctuations. Apart from $\alpha - \mu$ model that has been introduced previously, this chapter looks to model the irradiance fluctuations using the D-GG model. This statistical model is mathematically more involved in terms of complexity, but it can be exploited to represent all forms of atmospheric turbulence strength. Also, as considered in previous chapters, the pointing error, which occurs when the optical beam deviates from its intended direction due to misalignment or mechanical instability, results in signal loss or miscommunication. Addressing these challenges are crucial for realizing the full potential of FSO systems in practical deployments [1].

Furthermore, the integration of FSO systems with traditional RF counterparts has led to the emergence of mixed RF/FSO relay systems. These hybrid systems leverage the complementary strengths of RF and FSO technologies to enhance communication reliability and performance. While RF systems offer robustness against atmospheric conditions and obstacles, FSO links provide high bandwidth and low latency communication. By combining the two technologies, mixed RF/FSO relay systems can achieve improved coverage, throughput, and resilience to environmental challenges [2, 3].

The AF relaying strategy has garnered significant interest in the research community due to its simplicity and effectiveness in enhancing communication reliability. In AF relaying, the relay node amplifies and then forwards the received signal to the destination, thereby extending the communication range and improving signal strength. The amount of amplification required can either be constant or can be modified if the channel estimates are available on one hop. When the relay varies the gain that is to be provided to the incoming signal, it leads to variable gain relay. This approach is particularly well-suited for multiuser scenarios where simultaneous transmissions from multiple users can be leveraged to enhance system performance. User diversity (MUD) schemes have been proposed to exploit the diversity gain offered by simultaneous transmissions, where resources are allocated to users with the best channel conditions in various studies in the context of mixed RF/FSO relaying systems such as [4–14]. By dynamically selecting the user with the highest SNR for transmission, MUD schemes can maximize throughput and mitigate the effects of fading in RF channels [8]. When it comes to mixed RF/FSO relaying systems, using user diversity on the RF link also makes a lot of sense given that multiple RF users can be multiplexed together on to one large bandwidth FSO channel.

Moreover, the performance of practical wireless communication systems is often limited by interference from nearby transmissions. In relay networks, the presence of multiple CCIs can significantly degrade the quality of the desired signal at the receiving nodes. This form of interference occurs when multiple transmitters operate on the same frequency channel, leading to mutual interference and degradation of signal quality. Mitigating the impact of CCIs is essential for ensuring reliable communication in densely populated or interference-prone environments [15–19]. For the case of variable gain relaying systems, the CCIs can be present at both relay and destination nodes. Generally, even if the interference is included in analysis of mixed RF/FSO systems, for the ease of involved computations, it is assumed only at one of the nodes. It is important to consider the presence of interference at important nodes.

4.2.1 Literature

Extensive research efforts focusing on mixed RF/FSO relaying systems have been documented in the literature [2, 3, 8, 20, 21]. In [2], a dual hop mixed RF/FSO asymmetric relaying scheme was proposed, providing expressions for the outage probability of the overall system. The RF link was characterized using the Rayleigh distribution, while the FSO link was modeled considering Gamma-Gamma atmospheric turbulence. Similarly, in [3], researchers conducted a unified analysis of mixed RF/FSO relaying systems, considering pointing errors on the FSO link under both heterodyne detection and IM/DD. Performance analyses of dual hop RF/FSO links based on recently proposed generalized M distribution and D-GG fading models were presented in [8, 20, 21].

While many existing studies on FSO relay-assisted communications assume the absence of CCI, it is crucial to recognize that asymmetric mixed RF/FSO relay systems are inherently susceptible to CCI impacts due to the involvement of RF links. Additionally, frequency reuse in RF systems introduces additional interference sources [15]. Acknowledging the interference limited nature of mixed RF/FSO systems, authors in [16] derived expressions for outage probability and BER for an interference limited mixed RF/FSO AF relaying system. In this analysis, interfering signals have been modeled using Nakagami-m distribution fading, while FSO systems were assumed to follow the D-GG turbulence model, demonstrating exact and asymptotic performance of the overall system. Considering the impact of interference at the relay node, authors in [17] derived the closed-form expressions for outage probability. Furthermore, the analysis of a dual hop FSO/RF cooperative system, where FSO and RF links undergo Malaga and generalized-K distributed fading, has been reported in [18]. Recently, the effects of multiple CCIs on mixed RF/FSO systems were analyzed in [19], where authors derived expressions for outage probability, BER, and capacity with CSI assistance. Since then, a lot of works have been presented where interference has been included in analysis of mixed RF/FSO cooperative systems [12–14, 22–27].

Additionally, user diversity schemes have been explored in [5] within the context of wireless optical channels to assess the benefits of such configurations. In this study, the authors conducted latency calculations using various scheduling algorithms to evaluate the efficacy of user diversity schemes. Moreover, the impact of user diversity on FSO systems was investigated in [6], where a parallel RF-FSO relay strategy was compared with hybrid RF/FSO communications. The study derived expressions for the outage probability of the end-to-end system for different arrangement types, shedding light on the performance disparities between the two approaches. Furthermore, recent research in [8] delved into the analysis of mixed RF/FSO systems with user diversity, focusing on outage probability, BER, and capacity metrics. In this investigation, the RF channel was characterized by $\eta - \mu$ fading and assumed $\mathcal{M}$-distributed atmospheric turbulence was assumed for the FSO link. This comprehensive analysis provided insights into the performance enhancements achievable through user diversity in mixed RF/FSO systems. The research work [12] focused on providing the analysis of user diversity aided mixed RF/FSO cooperative relaying. The user diversity scheme has been modeled using the Nakagami-m statistical distribution. New PDF and CDF have been derived for Nakagami-m model that has been utilized to find the expression for outage probability and BER.

4.2.2 Motivation

Most previous studies have analyzed mixed RF/FSO relay systems under the assumption of interference-free transmission. Moreover, the few works that have addressed interference in FSO systems have not explored the potential improvement introduced by MUD schemes. Furthermore, there is a gap in the literature regarding the analysis of mixed RF/FSO systems with MUD and interference, considering a generalized distribution to model scintillation-based fading on the FSO link. Notably, the D-GG irradiance statistical model is deemed more accurate in modeling turbulence conditions compared to the Gamma-Gamma, M distribution, and double Weibull models [21]. We propose an asymmetric mixed RF/FSO variable gain relay scheme with SIM and assume MUD at the source node to enhance the performance of the RF link. Additionally, to make the system model more practical, we consider the effect of multiple CCIs on both the relay and destination nodes. Consequently, the system model presented in this chapter differs from those given in previous works [5, 6, 13–19, 21–28]. In fact, the present chapter works to utilize the model introduced in [12] and extend the analysis in a more comprehensive manner. Not only that, this chapter considers a different model to represent FSO turbulence as compared to [12], the analysis is more general where coding gain, diversity order, and ergodic capacity have been presented. Moreover, we model the RF link using the Nakagami-m fading model [29]. We consider unequal interferers with nonidentical statistical characteristics on both links to observe the effect of simultaneous transmission from different types of communication systems. Atmospheric turbulence on the FSO link is modeled using the D-GG fading distribution. Certainly, the various analysis metrics addressed in this chapter will provide a better understanding of the overall system performance due to the closer match of the distribution to actual fading conditions.

4.3 System and Channel Models

In this section, the system and channel models of the considered relaying strategy are discussed.

4.3.1 System Model

We consider an asymmetric mixed RF/FSO dual hop cooperative relaying system, where K users communicate with the destination node D through the relay node R. As the number of users on the RF link increases, the randomness offered by the RF channel increases, which is utilized by the node R for increasing system throughput [5, 6]. Moreover, the K users communicate with the relay node in the presence of L CCIs T the relay node. On source-to-relay link, the signal received $y_{1,j}$ from jth-user at node R is given by

$$y_{1,j} = \sqrt{P_{1,j}}h_{1,j}d_{RF} + \sqrt{P_{I1}}\sum_{i=1}^{L} h_{1,i}x_{r,i} + N_{RF} \tag{4.1}$$

where $h_{1,j}$ is the fading amplitude of RF channel carrying information d_{RF} of jth-user that follows Nakagami-m distribution. P_j stands for useful transmit power, whereas P_{I1} denotes the interference power with channel coefficient $h_{R,i}$ at the relay node R. The ith interferer at relay node is denoted by $x_{r,i}$.

In the second phase of communication, the relay node sends the processed version of the signal $y_{1,j}$ to the destination node D. In the presence of M CCIs at the destination node, the received signal is expressed as

$$\begin{aligned} y_{j,2} = {} & (\eta I)^{\frac{\varrho}{2}} G\sqrt{P_{1,j}}h_{1,j}d_{RF} + (\eta I)^{\frac{\varrho}{2}} G\sqrt{P_{I_1}}\sum_{i=1}^{L} h_{R,i}x_{r,i} \\ & + (\eta I)^{\frac{\varrho}{2}}\sqrt{P_{I_2}}\sum_{i=1}^{M} h_{D,i}x_{d,i} + N_o \end{aligned} \tag{4.2}$$

In (4.2), the overall noise is represented as $N_o = \eta I G N_{RF} + N_{FSO}$, where η is the electrical-to-optical conversion coefficient and I is the irradiance fluctuation on the FSO link, while G denotes the gain provided at the relay node. $x_{d,i}$ denote the modulation symbol of the ith interferer at node D with fading amplitude $h_{D,i}$. The channel attenuation of interferers on relay and destination nodes, $h_{R,i}$ and $h_{D,i}$, are assumed to follow Nakagami-m distributed fading. As mentioned in [15], the gain G can be formulated as

$$G^2 = \frac{1}{P_{1,j}|h_{1,j}|^2 + P_{I_1}\sum_{i=1}^{L}|h_{R,i}|^2 + \sigma_n{}^2} \tag{4.3}$$

For the cooperative relay strategy, the end-to-end SINR can be obtained using (4.2) as [15]

$$\gamma_{e2e} = \frac{\gamma_{1,j}^{o}\gamma_{2}^{o}}{\gamma_{1,j}^{o} + \gamma_{2}^{o} + 1} \tag{4.4}$$

where $\gamma_{1,j}^{o}$ is the effective instantaneous SINR on the RF link corresponding to the jth user considering both RF fading and the interference statistics and is given by the relation $\gamma_{1,j}^{o} = \frac{\gamma_k}{1+\sum_{i=1}^{L}\gamma_{I_r,i}}$. On the other hand, γ_{2}^{o} denotes the effective SINR on the wireless optical link considering atmospheric scintillation, pointing errors, path loss and interference defined as $\gamma_{2}^{o} = \frac{\gamma_2}{1+\sum_{i}^{M}\gamma_{I_d,i}}$. The variables γ_{I_r} and γ_{I_d} denote interfering signal at relay and destination nodes, respectively. For mathematical tractability, the analysis presented in the chapter considers an upper bound, $\gamma_{e2e} < \gamma_{up}$, such that

$$\gamma_{e2e} = \min\left(\gamma_{1,j}^{o}, \gamma_{2}^{o}\right) \tag{4.5}$$

4.3.2 *Channel Models*

For the purpose of analysis, the various channel models involved in the system model have been defined. The RF link has been assumed to undergo slow fading that is represented by the Nakagami-*m* fading model. The closed-form expressions for the multiuser links have been derived. We also model the optical wireless link using the D-GG model.

4.3.2.1 MUD RF Link

Under the assumption of Nakagami-*m* distribution, the fading coefficients on the RF link, for the kth user, follow the PDF given by [29]

$$f_{\gamma_k}(\gamma) = \frac{\beta^{m_1}}{\Gamma(m_1)}\gamma^{m_1-1}\exp(-\beta\gamma) \tag{4.6}$$

where m_1 is the shape parameter and $\beta = \frac{m_1}{\bar{\gamma}_k}$. Furthermore, the PDF of instantaneous SNR on the RF link with K users can be obtained using the definition of ordered statistics as

$$f_{\gamma_{1,j}}(\gamma) = K[F_{\gamma_k}(\gamma)]^{K-1} f_{\gamma_k}(\gamma) \tag{4.7}$$

Hence, considering (4.6) and (4.7), the closed-form expression for the PDF of RF link experiencing Nakagami-*m* fading for integer values of fading parameter m_1 can be obtained as [12]

$$f_{\gamma_1,j}(\gamma) = \frac{K\beta^{m_1}}{\Gamma(m_1)} \sum_{n_1=0}^{K-1} (-1)^{n_1} \binom{K-1}{n_1} \times \sum_{n_2=0}^{n_1(m_1-1)} \zeta_{n_1 n_2}(m_1) \gamma^{m_1+n_2-1} \beta^{n_2} \exp(-\beta(n_1+1)\gamma) \tag{4.8}$$

where $\zeta_{n_1n_2}(m_1)$ is the coefficient of multinomial expansion involved in the derivation of (4.8) which can be recursively obtained using the relation $\zeta_{n_1n_2} = \sum_{b=n_1-x+1}^{n_1} \frac{\lambda_{bn_2-1}}{n_1-b} \mathrm{I}_{[0,(n_2-1)(x-1)]}$, where $I_{[a,b]}$ is the indicator function as defined in [29, Eq. (9.120)]. Moreover, the PDF of the total interference-to-noise ratio (INR) $\sum_{i=1}^{L} \gamma_{I_r,i}$ can be expressed as

$$f_{I_r}(\gamma) = \left[\frac{m_1}{\Omega_{I1}}\right]^{m_1 L} \frac{\gamma^{m_1 L-1}}{\Gamma(m_1 L)} \exp\left(-\frac{m_1 L}{\Omega_{I1}}\gamma\right) \tag{4.9}$$

where m_1 is Nakagami-m fading parameter and Ω_{I1} is the average INR on the RF link. Given that γ_1 and γ_{I_r} are independent, the PDF of $\gamma_{1,j}^{o}$ is given by

$$f_{\gamma_1,j}^{o}(\gamma) = \int_0^\infty (1+y) f_{\gamma_1}[(1+y)\gamma] f_{I_r}(y) dy \tag{4.10}$$

Substituting (4.8) and (4.9) in (4.10), the PDF can be expressed as

$$\gamma_{1,j}^{o} = \sum_{n_1=0}^{K-1} (-1)^{n_1} \binom{K-1}{n_1} \sum_{n_2=0}^{n_1(m_1-1)} \zeta_{n_1n_2}(m_1) \gamma^{m_1+n_2-1} \beta^{n_2} e^{[-\beta(n_1+1)\gamma]} \times \int_0^\infty (1+y)^{m_1+n_2-1} y^{m_1 L-1} e^{-[\beta(n_1+1)\gamma + \frac{m_1}{\Omega_{I1}}]y} dy,$$

where $A_1 = \frac{K\beta^{m_1}}{\Gamma(m_1 L)\Gamma(m_1)} \left(\frac{m_1}{\Omega_{I_1}}\right)^{m_1 L}$. Applying [30, Eq. (383.5)], the objective PDF can be simplified as [12]

$$f_{\gamma_1,j}^{o}(\gamma) = \sum_{n_1=0}^{K-1} \sum_{n_2=0}^{n_1(m_1-1)} A_2 \gamma^{m_1+n_2-1} e^{[-\beta(n_1+1)\gamma]} U\big(m_1 L, m_1 L + 1 + m_1 + n_2, \mathcal{B}_0(\gamma)\big) \tag{4.11}$$

where $U(a, b, z)$ is the confluent hypergeometric function of second kind as defined in [30, Eq. (9.211.4)], $\mathcal{B}_0(\gamma) = \beta(n_1+1)\gamma + \frac{m_1}{\Omega_{I1}}$, the constant $A_2 = A_1(-1)^{n_1}\binom{K-1}{n_1}\Gamma(m_1 L)\zeta_{n_1n_2}(m_1)\beta^{n_2}$. The cumulative distribution function (CDF) on the RF link undergoing Nakagami-m fading in the presence of CCIs can be obtained using $F^o_{\gamma_1}(\gamma) = \int_0^\infty f^o_{\gamma_1}(\gamma)d\gamma$. After some mathematical manipulations, the CDF of $\gamma^o_{1,j}$ can be expressed as follows [12]:

$$F^o_{\gamma_1,j}(\gamma) = \frac{K}{\Gamma m_1} \sum_{n_1=0}^{K-1} \sum_{n_2=0}^{n_1(m_1-1)} A_3 \Bigg[1 - \sum_{m=0}^{n_2+m_1-1} e^{(-\beta(n_1+1)\gamma)} \frac{\{(n_1+1)\beta\gamma\}^m}{m!} \times U\Big(m_1 L, m_1 L + 1 + m, \mathcal{B}_0(\gamma) \Big) \Bigg] \tag{4.12}$$

where A_3 is given by

$$A_3 = (-1)^{n_1} \binom{K-1}{n_1} (m_1 + n_2 - 1)! \zeta_{n_1 n_2}(m_1)(n_1+1)^{-n_2-m_1} \tag{4.13}$$

4.3.2.2 FSO Link

On the other hand, it is assumed that $R \to D$ link encompasses the turbulence-induced fading I_a and the pointing errors I_p such that $I = I_a I_p$. The pointing error jitter is given as [1]

$$I_p = A_0 \exp\left(-\frac{2R^2}{\omega_{eq}^2} \right) \tag{4.14}$$

where the pointing error parameters are defined again as $\omega_{eq}^2 = \omega_z^2 \frac{\sqrt{\pi}\,\mathrm{erf}(v)}{2v\exp(-v^2)}$ is the equivalent beam width, $v = \sqrt{\frac{\pi}{2}} \frac{d}{\omega_z}$ and $A_0 = [\mathrm{erf}(v)]^2$. Here z is beam waist (calculated at e^2) at distance z, while d is the detection radius of the FSO aperture. The atmospheric turbulence fading I_a on the FSO link is modeled as double generalized gamma (GG) fading model [21]. The irradiance $I_a = I_x I_y$, such that $I_x \sim GG(\alpha_1, \beta_1, \Omega_1)$ and $I_y \sim GG(\alpha_2, \beta_2, \Omega_2)$. β_1 and β_2 are shaping parameters, whereas $\alpha_1, \Omega_1, \alpha_2, \Omega_2$ are calculated based on the variances of the small- and large-scale fluctuations. The PDF of I_a can be expressed as [21]

$$f(I_a) = \frac{y\sigma^{\beta_1-\frac{1}{2}}\lambda^{\beta_2-\frac{1}{2}}(2\pi)^{1-\frac{\sigma+\lambda}{2}}}{\Gamma(\beta_1)\Gamma(\beta_2)I_a} \mathrm{G}^{0,\lambda+\sigma}_{\lambda+\sigma,0}\left[\left(\frac{\Omega_2}{I_a^{\alpha_2}}\right)^{\lambda} \frac{\lambda^\lambda \sigma^\sigma \Omega_1}{\beta_1^\sigma \beta_2^\lambda} \middle| \begin{matrix} 1-\tau_0 \\ - \end{matrix} \right] \tag{4.15}$$

where $\tau_0 = [\Delta(\sigma : \beta_1), \Delta(\lambda : \beta_2)]$ with $\Delta(z : x)$ representing $[\frac{x}{z}, \frac{x+1}{z}, \ldots, \frac{x+z-1}{z}]$, where $\mathrm{G}^{m,n}_{p,q}\left[\cdot\middle|\ \right]$ is the Meijer-G function defined in [30, Eq. (9.301)] and $y = \alpha_2\lambda$. Moreover, the SNR PDF on FSO link experiencing a double GG fading model with the impact of pointing errors is given by [21]

$$f_{\gamma_2}(\gamma) = \frac{\mathcal{D}_1}{r\gamma} \mathrm{G}^{\lambda+\sigma+1,0}_{1,\lambda+\sigma+1}\left[\mathcal{D}_2 z^y \left(\frac{\gamma}{\mu_r}\right)^{\frac{y}{r}} \middle| \begin{matrix} \tau_2 \\ \tau_1 \end{matrix}\right] \tag{4.16}$$

with

$$\mathcal{D}_1 = \frac{\xi^2 \sigma^{\beta_1 - \frac{1}{2}} \sigma^{\beta_2 - \frac{1}{2}} (2\pi)^{1-\frac{\sigma+\lambda}{2}}}{\Gamma(\beta_1)\Gamma(\beta_2)} \tag{4.17}$$

$$\mathcal{D}_2 = \frac{\beta_1^{\sigma} \beta_2^{\lambda}}{\lambda^{\lambda} \sigma^{\sigma} \Omega_1^{\sigma} \Omega_2^{\lambda}} \tag{4.18}$$

where $\tau_1 = \left[\frac{\xi^2}{y}, \Delta(\sigma : \beta_1), \Delta(\lambda : \beta_2)\right]$, $\tau_2 = \left[1 + \frac{\xi^2}{y}\right]$, $\mu_r = \frac{(\eta \mathbb{E}(I))^r}{N_0}$, and r denotes the kind of demodulation scheme [21]. Considering $\mathcal{D}_3 = \prod_{g=1}^{\sigma+\lambda} \Gamma\left(\frac{1}{y} + \tau_{0,g}\right)$, z can be given as $z = \frac{\mathcal{D}_1 (\mathcal{D}_2)^{1/y} \mathcal{D}_3}{(1+\xi^2)}$.

Furthermore, the unified CDF of FSO channel over double GG atmospheric turbulence with pointing errors can be formulated as follows [21]:

$$F_{\gamma_2}(\gamma) = \mathcal{D}_4 \mathrm{G}^{n,1}_{r+1,n+1}\left[\mathcal{D}_5 \left(\frac{\gamma}{\mu_r}\right)^{y} \middle| \begin{matrix} 1, \tau_3 \\ \tau_4, 0 \end{matrix}\right] \tag{4.19}$$

where $n = r(\lambda + \sigma + 1)$, $\mathcal{D}_4 = \frac{\xi^2 \sigma^{\beta_1 - \frac{1}{2}} \sigma^{\beta_2 - \frac{1}{2}} (2\pi)^{1-\frac{r(\sigma+\lambda)}{2}} r^{(\beta_1+\beta_2-2)}}{y\Gamma(\beta_1)\Gamma(\beta_2)}$, $\mathcal{D}_5 = \left(\frac{\mathcal{D}_2 z^y}{r^{\sigma+\lambda}}\right)^r$, $\tau_3 = [\Delta(r : \tau_2)]$ and $\tau_4 = [\Delta(r : \tau_1)]$ comprising of $r(\lambda + \sigma + 1)$ terms. Moreover, the interference on the destination is assumed to follow the PDF [28]

$$f_{I_d}(\gamma) = \frac{\gamma^{M-1}}{\Omega_{I2}(M-1)!} \exp\left(-\frac{\gamma}{\Omega_{I2}}\right) \tag{4.20}$$

where Ω_{I2} is the INR on the FSO link. Given that I_d and γ_2 are independent, the CDF of γ_2^o can be derived using

$$F^o_{\gamma_2}(\gamma) = \int_0^{\infty} F_{\gamma_2}(\gamma(1+x)) f_{I_d}(x) dx \tag{4.21}$$

Substituting (4.19) and (4.20) in (4.21), and assuming $1 + y = z$, above integral may be formulated as

$$F^o_{\gamma_2}(\gamma) = \mathcal{D}_6 (\mathcal{J}_1 - \mathcal{J}_2) \tag{4.22}$$

where $\mathcal{D}_6 = \frac{\mathcal{D}_4}{{\Omega_{I2}}^M (M-1)!} \exp(1/\Omega_{I2})$ and

$$\mathcal{J}_1 = \int_0^{\infty} (z-1)^{M-1} e^{-\frac{z}{\Omega_{I2}}} \mathrm{G}^{n,1}_{r+1,n+1}\left[\mathcal{D}_5 \left(\frac{z\gamma}{\mu_r}\right)^{y} \middle| \begin{matrix} 1, \tau_3 \\ \tau_4, 0 \end{matrix}\right] \tag{4.23}$$

Resorting to binomial expansion $(z-1)^{M-1} = \sum_{i=0}^{M-1}\binom{M-1}{i}(-1)^{M-1-i}z^i$ as per [30, Eq. (1.11)] for integer values of M and with the aid of [31, Eq. (07.34.21.0013.01)], integral $\mathcal{J}_1$ can be given as

$$\mathcal{J}_1 = \sum_{i=0}^{M-1} x_i \mathcal{D}_{10} \mathrm{G}_{r+y+1,n+1}^{n,y+1}\left[\mathcal{D}_7(\gamma)^y \middle| \begin{matrix} 1,\tau_3,\tau_5 \\ \tau_4,0 \end{matrix}\right] \tag{4.24}$$

where $\mathcal{D}_7 = \mathcal{D}_5(y\Omega_{I2}/\mu_r)^y$, $\mathcal{D}_{10} = \frac{y^{i+\frac{1}{2}}(\Omega_{I2})^{i+1}}{(2\pi)^{\frac{y-1}{2}}}$, and $\tau_5 = \left[\frac{(-i)}{y}, \ldots, \frac{(v-1-i)}{y}\right]$.

On the other hand, assuming $x_i = \binom{M-1}{i}(-1)^{M-1-i}$, the integral $\mathcal{J}_2$ can be given as

$$\mathcal{J}_2 = \sum_{i=0}^{M-1} x_i \int_0^1 (z)^i \quad e^{-\frac{z}{\Omega_{I2}}} \mathrm{G}_{r+1,n+1}^{n,1}\left[\mathcal{D}_5\left(\frac{z\gamma}{\mu_r}\right)^y \middle| \begin{matrix} 1,\tau_3 \\ \tau_4,0 \end{matrix}\right] \tag{4.25}$$

The above integral $\mathcal{J}_2$ can be expressed in complex integral form on using [31, Eq. (07.34.02.0001.01)] as illustrated in (4.26) below

$$\mathcal{J}_2 = \sum_{i=0}^{M-1} \frac{x_i}{(2\pi i)^2} \int_{\mathcal{L}_1}\int_{\mathcal{L}_2} \Gamma(s_1) \frac{\Gamma(-s_2)\prod_{j=1}^{n}\Gamma(\tau_{4,j}+s_2)}{\Gamma(1-s_2)\prod_{j=2}^{r+1}\Gamma(s_2+\tau_{3,j})} \left(\frac{1}{\Omega_{I_2}}\right)^{-s_1} \left(\gamma^y \mathcal{D}_5\right)^{-s_2}$$
$$\times \left[\int_0^1 z^{i-s_1-s_2\frac{\alpha}{\rho}}(1-z)^{\omega-1}dz\right] ds_1 ds_2 \tag{4.26}$$

Using [30, Eq. (8.380)] and [32, Eq. (28–30)] and with some mathematical manipulations, the integral $\mathcal{J}_2$ in (4.26) can be evaluated as shown in (4.27).

$$\begin{aligned}\mathcal{J}_2 &= \sum_{i=0}^{M-1} x_i \mathrm{H}_{1,1:0,1:r+1,n+1}^{0,1:1,0:n,1} \\ &\quad\times \left[\begin{matrix} (-i:1,v) \\ (-i-1:1,v) \end{matrix} \middle| \begin{matrix} - \\ (0,1) \end{matrix} \middle| \begin{matrix} (1,1),(\tau_3,[1]_{\tau_3}) \\ (\tau_4,[1]_{\tau_4}),(o,1) \end{matrix} \middle| \frac{1}{\Omega_{I2}}, \mathcal{D}_5\left(\frac{z\gamma}{\mu_r}\right)^y\right] \\ &= \sum_{i=0}^{M-1} x_i \mathcal{H}\left[\Omega_{I2}, \left(\frac{\gamma}{\mu_r}\right)^y\right]\end{aligned} \tag{4.27}$$

In the expression given in (4.27), we have used $[1]_x$ to represent sequence of $1^{'s}$ of length x and $\mathrm{H}_{p_1,q_1:p_2,q_2:p_3,q_3}^{0,m_1:m_2,n_2:m_3,n_3}[\mathcal{A}_1,\mathcal{A}_2] = \mathcal{H}[\mathcal{A}_1,\mathcal{A}_2]$ is the bivariate Fox-H function given in [33, Eq. (28)]. Moreover, the PDF of end-to-end SINR can be derived by utilizing the relationship (4.10). Using the identity [31, Eq. (07.34.21.0013.01)] and [33], the closed-form expression for the overall PDF can be written as shown below

$$f^{o}_{\gamma_2}(\gamma)$$

$$= \mathcal{D}_{11} \sum_{i=0}^{M-1} x_i \left[\mathcal{D}_{12} \mathrm{G}^{r(\lambda+\sigma+1),y}_{r+y,r(\lambda+\sigma+1)} \left[\mathcal{D}_8 \gamma^{y} \middle| \begin{smallmatrix} \tau_6 \\ \tau_7 \end{smallmatrix} \right] - \mathcal{H}\left[\Omega_{I2}, \mathcal{D}_2 z^{y} \left(\frac{\gamma}{\mu_r} \right)^{y/r} \right] \right] \tag{4.28}$$

In (4.28), the parameters involved in Meijer-G function have been derived as $\tau_6 = \left[\frac{\tau_2}{r}, \ldots, \frac{\tau_2+r-1}{r}, \frac{1-(i+2)}{y}, \ldots, \frac{y-(i+2)}{y}\right]$ and $\tau_7 = \left[\frac{\tau_1}{r}, \ldots, \frac{\tau_1+r-1}{r}, \ldots, \frac{\tau_1(\lambda+\sigma+1)+r-1}{r}\right]$. The argument of the Meijer-G equals $\mathcal{D}_8 = \mathcal{D}_2 z^{y} \{r\}^{(-r(\lambda+\sigma))} \left(\frac{\Omega_I 2}{y\mu_r}\right)^{y}$, $\mathcal{D}_{11} = \frac{\mathcal{D}_1 \exp(\frac{1}{\Omega_{I2}})}{r\lambda \Omega_{I2}{}^{M}(M-1)!}$, and $\mathcal{D}_{12} = \frac{y^{i+\frac{3}{2}} (\Omega_{I2})^{i+2} (r)^{c*}}{(2\pi)^{\left\{y-1+\left(\frac{(r-1)(\lambda+\sigma)}{2}\right)\right\}}}$. The bivariate Fox's H function is defined in (4.30). Substituting (4.24) and (4.27) into (4.22), we obtain the closed-form equation of CDF on FSO link in the presence of multiple interferers as given below

$$F^{o}_{\gamma_2}(\gamma) = \mathcal{D}_6 \sum_{i=0}^{M-1} x_i \mathcal{D}_{10} \mathrm{G}^{n,y+1}_{r+y+1,n+1} \left[\mathcal{D}_7 (\gamma)^{y} \middle| \begin{smallmatrix} 1,\tau_3,\tau_5 \\ \tau_4, 0 \end{smallmatrix} \right] - \mathcal{H}_2 \left[\Omega_{I2}, \left(\frac{\gamma}{\mu_r} \right)^{y} \right] \tag{4.29}$$

where

$$\mathrm{H}^{0,1:1,0:\lambda+\sigma+1,1}_{1,1:0,1:1,\lambda+\sigma+1} \left[\begin{matrix} (-i-1:1,\, y/r) \\ (-i-2:1,\, y/r) \end{matrix} \middle| \begin{matrix} \text{—} \\ (0,1) \end{matrix} \middle| \begin{matrix} (\tau_2, [1]_{\tau_2}) \\ (\tau_1, [1]_{\tau_1}) \end{matrix} \middle| \mathcal{A}, \mathcal{B} \right] = \mathcal{H}_2 [\mathcal{A}, \mathcal{B}] \tag{4.30}$$

4.4 Performance Analysis

In this section, we investigate various performance metrics of relay system based on the aforementioned SNR bound.

4.4.1 *Exact Outage Probability*

The outage probability is an important performance indicator of wireless communication system. Outage probability is defined as the probability that the instantaneous SNR falls below a predefined threshold γ_{th}. The end-to-end CDF $F_{e_2e}(\gamma)$ of the overall system can be written as

$$F_{e_2e}(\gamma) = F^{o}_{\gamma_1,j}(\gamma) + F^{o}_{\gamma_2}(\gamma) - F^{o}_{\gamma_1,j}(\gamma) F^{o}_{\gamma_2}(\gamma) \tag{4.31}$$

Hence, substituting (4.12) and (4.29) into (4.31), and by putting $\gamma = \gamma_{th}$, we can obtain the overall CDF of the considered system.

4.4.2 Asymptotic Outage Probability

The exact expression derived in (4.31) fails to offer quick insights into the performance of overall system. A simpler asymptotic (high SNR) expression for outage probability can be developed to get more insights about system's performance. It is worth noting that Meijer-G and/or Fox-H functions are given as complex integrals, which can be computed numerically only. This computation can be subject to underflow given the fact that the arguments of these functions become arbitrarily small as the average SNR increases. Here, we focus on expression of outage probability for the case when average SNR on RF and FSO link tend to infinity, i.e., $\bar{\gamma}_k \to \infty$ and $\mu_r \to \infty$. The CDF of Nakagami-m distributed jth-user can be written as $F_{\gamma_k}(\gamma) = 1 - \Gamma\left(m_1, m_1\left(\frac{\gamma}{\bar{\gamma}_k}\right)^{m_1}\right)$, where $\Gamma(x, y)$ is the upper incomplete Gamma function defined in [30, Eq. (8.310.1)]. For integer values of m_1, we note that $\Gamma\left(m_1, m_1\left(\frac{\gamma}{\bar{\gamma}_k}\right)^{m_1}\right) = \exp\left(-m_1\left(\frac{\gamma}{\bar{\gamma}_k}\right)^{m_1}\right)\sum_{j=0}^{m_1-1}\frac{\{m_1(\frac{\gamma}{\bar{\gamma}_k})^{m_1}\}^j}{!j}$. Moreover, as $\bar{\gamma}_k \to \infty$, the upper incomplete gamma function will be dominated by $j = 0$ term and therefore can be approximated as $\Gamma\left(m_1, m_1\left(\frac{\gamma}{\bar{\gamma}_k}\right)^{m_1}\right) \simeq \exp\left(-m_1\left(\frac{\gamma}{\bar{\gamma}_k}\right)^{m_1}\right)$. Upon utilizing the Taylor series representation of the exponential term in the CDF $F_{\gamma_k}(\gamma) \simeq 1 - \left[1 - m_1\left(\frac{\gamma}{\bar{\gamma}_k}\right)^{m_1} + \frac{m_1^2}{2!}\left(\frac{\gamma}{\bar{\gamma}_k}\right)^{2m_1} - \ldots\right]$. Furthermore, as $\bar{\gamma}_k \to \infty$, we observe that only the first-order terms in series remain significant, and the CDF may be simplified as $F_{\gamma_k}(\gamma) \simeq m_1\left(\frac{\gamma}{\bar{\gamma}_k}\right)^{m_1}$. The asymptotic expression of the effective CDF on the RF link $F^o_{\gamma_1}(\gamma)$ in the presence of multiple interferers may be computed as

$$F^o_{\gamma_1}(\gamma) \simeq A_4 \gamma^{Km_1} \int_0^\infty (\gamma(1+x))^{m_1} f_{I_r}(x)dx \tag{4.32}$$

where $A_4 = \frac{\{m_1\}^K}{\bar{\gamma}_k}\left[\frac{m_1}{\Omega_{I1}}\right]^{m_1 L}$. On substituting (4.9), into (4.32), with some mathematical manipulations, the corresponding approximate CDF expression can be stated as

$$F^o_{\gamma_1,j}(\gamma) \simeq \left(\frac{\Lambda_1 \bar{\gamma}_k}{\gamma}\right)^{-Km_1} \tag{4.33}$$

where $\Lambda_1 = m_1^{\frac{-1}{m_1}} \sum_{j=0}^{Km_1}\left(\frac{m_1 L}{\Omega_{I_1}}\right)^{-j} \Gamma(j + m_1 L)$.

Moreover, for the FSO system analysis as $\mu_r \to \infty$, the argument of Meijer-G in (4.24) becomes very small, i.e., $\mathcal{D}_7 \to 0$. For these approximations, with the aid of [31, Eq. (07.34.06.0006.01)], $\mathcal{J}_1$ can be presented in terms of elementary functions as

$$\begin{aligned}\mathcal{J}_1 &\simeq \sum_{i=0}^{M-1} x_i \mathcal{D}_{10} \sum_{j=1}^{n} \left(\mathcal{D}_7 \gamma^y\right)^{\tau_{4,j}} \frac{\prod_{k=1}^{n} \Gamma\left(\tau_{4,k} - \tau_{4,j}\right)}{\tau_{4,j}} \frac{\prod_{k=2}^{y} \Gamma\left(1 - \tau_{5,k} + \tau_{4,j}\right)}{\prod_{k=y+2}^{y+1+r} \Gamma\left(\tau_{3,k} - \tau_{4,j}\right)} \\ &= \sum_{i=0}^{M-1} x_i \frac{y^i + \frac{1}{2}(\Omega_{I2})^{i+1}}{(2\pi)^{\frac{y-1}{2}}} \sum_{j=1}^{n} \left(\mathcal{D}_7 \gamma^y\right)^{\tau_{4,j}} \Lambda_3 \prod_{k=2}^{y} \Gamma\left(1 - \tau_{5,k} + \tau_{4,j}\right)\end{aligned} \tag{4.34}$$

where $\Lambda_3 = \frac{\prod_{k=1}^{n} \Gamma(\tau_{4,k} - \tau_{4,j})}{\tau_{4,j} \prod_{k=y+2}^{y+1+r} \Gamma(\tau_{3,k} - \tau_{4,j})}$. Additionally, the above given results will be further dominated by $\min(\tau_{4,j})$. Invoking [34, Theorem 1.7 and Theorem 1.11], the asymptotic expression for $\mathcal{J}_2$ can be formulated by evaluating the residue of the complex integration at the poles closest to the contour of integration. As per [34, Theorem 1.7 and Theorem 1.11], the asymptotic approximation could be accomplished by evaluating residues of the multivariate Fox's H function at the largest negative pole of $\Gamma(b_j + B_i s)$, $j = 1, \ldots, k$. Considering **s** to be largest negative pole of τ_4 along with the identity [30, Eq. (9.113)], and with the aid of [34, Eq. (1.2)], the asymptotic approximation of $\mathcal{J}_2$ can be derived as

$$\mathcal{J}_2 = \sum_{i=0}^{M-1} x_i \frac{\Gamma(1 + i + y\mathbf{s})}{\Gamma(2 + i + y\mathbf{s})} \left(\frac{\mathcal{D}_5 \gamma^y}{\mu_r^y}\right)^{\mathbf{s}} \Lambda_{31} F_1\left(1 + i + y\mathbf{s}, 2 + i + y\mathbf{s} : -\frac{1}{\Omega_{I2}}\right) \tag{4.35}$$

In (4.35), the simplicity of results is evident as the expression consists of all elementary functions only. Generally, the evaluation of Meijer-G and/or Fox's H function is performed by utilizing numerical integration in software packages. The obtained expansion in (4.35) upon considering only the dominant pole is adequate for most applications. Noting the fact that (4.31) can be approximated as $F_{e_2e}(\gamma) \simeq F^o_{\gamma_{1,j}}(\gamma) + F^o_{\gamma_2}(\gamma)$, where the negative term in (4.31) is omitted as it becomes minor at high SNR values and (4.31) is dominated by the first two terms only. Applying (4.34) and (4.35) in this, the high SNR approximation of end-to-end CDF can be achieved.

$$P_{out}^{\infty} \underset{\bar{\gamma}_k, \mu_r \to \infty}{\simeq} \left(\frac{\Lambda_1 \bar{\gamma}_k}{\gamma_{th}}\right)^{-Km_1} + \sum_{i=0}^{M-1} x_i \left(\frac{\Lambda_2 \mu_r}{\gamma_{th}}\right)^{-y\mathbf{s}} \tag{4.36}$$

where Λ_2 is defined in (4.37) as

$$\Lambda_2 = \left[\mathcal{D}_{10} \left[\mathcal{D}_5 (y \Omega_{I_2})^y \right]^{\mathbf{s}} \Lambda_3 \prod_{k=2}^{y} \Gamma \left(1 - \tau_{5,k} + \mathbf{s} \right) - \frac{\Gamma(1 + i + y\mathbf{s})}{\Gamma(2 + i + y\mathbf{s})} (\mathcal{D}_5)^{\mathbf{s}} \Lambda_{31} \right.$$
$$\left. \times F_1 \left(1 + i + y\mathbf{s}, 2 + i + y\mathbf{s} : -\frac{1}{\Omega_{12}} \right) \right]^{\frac{-1}{y\mathbf{s}}} \tag{4.37}$$

At high SNR values, the outage probability of a certain transmission link can be $P_{out}^{\infty} \simeq (G_c \bar{\gamma}_k)^{-G_d}$, where G_c is the coding gain, whereas G_d is the diversity order of the overall system [35]. In order to concrete results further, two separate scenarios are discussed as follows:

Case I *Weak Turbulence*: The scenario when the atmospheric scintillation is small, as $\mu_r \rightarrow \infty$, the CDF of second hop SNR $F_{\gamma_2}^{o}(\gamma)$ diminishes quickly. Therefore, the RF link dominates the overall performance.

$$P_{out}^{\infty} \underset{\gamma_k, \mu_r \rightarrow \infty}{\simeq} \left(\frac{\Lambda_1 \bar{\gamma}_k}{\gamma_{th}} \right)^{-K m_1} \tag{4.38}$$

It can be easily observed that the system at weak turbulence conditions can achieve a coding gain of $G_c = \frac{\Lambda_1}{\gamma_{th}}$, while the diversity order of the system can be obtained as $G_d = K m_1$, and therefore the diversity order is determined by the number of users of the multiuser diversity system.

Case II *Strong Turbulence*: This makes sense, as under strong scintillation conditions and large pointing errors, the second hop channel will be dominating the system performance. In such a case, as $\bar{\gamma}_k \rightarrow \infty$, the $F_{\gamma_1}^{o}(\gamma) \rightarrow 0$. The OP can be represented as

$$P_{out}^{\infty} \underset{\gamma_k, \mu_r \rightarrow \infty}{\simeq} \sum_{i=0}^{M-1} x_i \left(\frac{\Lambda_2 \mu_r}{\gamma_{th}} \right)^{-y\mathbf{s}} \tag{4.39}$$

It can be observed from (4.39) that the diversity order of the studied system over strong turbulence on second hop channel is $G_d = \min(y\tau_4) = y\mathbf{s}$. It can be concluded that the diversity order depends upon pointing error, type of demodulation scheme, and the parameters of double GG channel model. On the other hand, the coding gain can be given as $G_c = \frac{\Lambda_2}{\gamma_{th}}$.

4.4.3 Optimum Power Allocation

In this section, a transmission power allocation scheme for the multiuser RF/FSO system under consideration is discussed. A total power budget of P_T is to be allocated optimally among the RF and FSO transmitting nodes during one complete round of data transmission. We intend to minimize the asymptotic expression of outage probability in (4.36) by restricting the total power P_T of the system. The high SNR approximation in (4.36) can be rewritten as

$$P_{out}^{\infty} \simeq \Lambda_4 P_{_{1,j}}{}^{a_0} + \sum_{i=0}^{M-1} x_i \Lambda_5 P_{_{FSO}}{}^{a_1} \tag{4.40}$$

where $\Lambda_4 = (\Lambda_1/N_{_{RF}}\gamma_{th})^{a_0}$ with $a_0 = -(Km_1)$ and $\Lambda_5 = (\Lambda_2/N_{_{FSO}}\gamma_{_{th}})^{a_1}$ with $a_1 = -ys$. Assuming $P_{_{1,j}}{}^*$, $P_{_{1,j}}{}^*$ to be optimum powers on the RF and FSO link, respectively, we formulate the objective function $\bar{\mathcal{F}}$ as given below

$$\bar{\mathcal{F}} = (P_{_{1,j}}{}^*, P_{_{1,j}}{}^*) = \min_{P_{_{1,j}}, P_{_{1,j}}} \arg P_{out}^{\infty} \tag{4.41}$$

By taking the first-order derivative of $\bar{\mathcal{F}}$ with respect to $P_{_{1,j}}$, we get

$$\frac{d\bar{\mathcal{F}}}{dP_{_{1,j}}} = a_1 \Lambda_4 P_{_{1,j}}{}^{a_0-1} - \sum_{i=0}^{M-1} x_i a_1 \Lambda_5 (P_T - P_{_{1,j}})^{a_1-1} \tag{4.42}$$

Equating the above optimization of objective function result to 0, we have

$$P_{_{1,j}}{}^* = \left[\left(P_T - P_{_{1,j}}{}^*\right)^{a_1-1} \frac{\sum_{i=0}^{M-1} x_i a_1 \Lambda_2}{a_0 \Lambda_1} \right]^{\frac{1}{a_0-1}} \tag{4.43}$$

A closed-form solution for calculating $P_{_{1,j}}$ is tough to obtain from the above results. Nevertheless, the numerical solution can be found by standard iterative methods such as Bisection's method or Newton's method.

4.4.4 Exact Bit Error Rate

The BER of the wireless fading channel gives important implications about the overall reliability of the network. Although the assumed system is complex, this section deals with providing an exact closed-form analytical expression for the error rate. It can be noted that BER for a variety of binary modulation schemes can be evaluated using [36]

$$\bar{P}_b = \frac{q^p}{2\Gamma(p)} \int_0^{\infty} \gamma^{p-1} \exp(-q\gamma) F_{e_2e}(\gamma) d\gamma \tag{4.44}$$

In order to derive the analytical expression of BER, the end-to-end CDF $F_{e_2e}(\gamma)$ of the dual hop scheme in (4.31) is substituted into (4.44). The resulting BER can be further expressed as follows:

$$P_b = P_{b1} + P_{b2} - \frac{K}{\Gamma m_{RF}} \sum_{n_1=0}^{K-1} \sum_{n_2=0}^{n_1(m_{RF}-1)} \left[P_{b2} - P_{b32} + P_{b33} \right] \tag{4.45}$$

The BER integrals that need to be solved can be defined as follows:

$$P_{b1} = \frac{K}{\Gamma m_1} \sum_{n_1=0}^{K-1} \sum_{n_2=0}^{n_1(m_1-1)} \frac{q^p A_3}{2\Gamma(p)} \Bigg[\int_0^\infty \gamma^{p-1} e^{-q\gamma} d\gamma - \sum_{m=0}^{n_2+m_1-1} \frac{\{(n_1+1)\beta\}^m}{m!} \times \underbrace{\int_0^\infty \gamma^{m+p-1} \frac{U\Big(m_1 L, m_1 L + 1 + m, \mathcal{B}_0(\gamma)\Big)}{e^{\{q+\beta(n_1+1)\}\gamma}} d\gamma}_{\mathcal{J}_3} \Bigg] \tag{4.46}$$

$$P_{b2} = \mathcal{D}_6 \sum_{i=0}^{M-1} x_i \Bigg\{ \mathcal{D}_{10} \underbrace{\int_0^\infty \frac{\gamma^{p-1}}{e^{q\gamma}} \mathrm{G}_{r+y+1,n+1}^{n,y+1} \left[\mathcal{D}_7 (\gamma)^y \Big| \begin{smallmatrix} 1,\tau_3,\tau_5 \\ \tau_4,0 \end{smallmatrix} \right] d\gamma}_{\mathcal{J}_4} - \underbrace{\int_0^\infty \gamma^{p-1} e^{-q\gamma} \mathcal{H}\left[\Omega_{I2}, \left(\frac{\gamma}{\mu_r}\right)^y \right] d\gamma}_{\mathcal{J}_5} \Bigg\} \tag{4.47}$$

$$P_{b32} = \sum_{m=0}^{n_2+m_1-1} \sum_{i=0}^{M-1} x_i \mathcal{D}_6 \mathcal{D}_{10} \underbrace{\int_0^\infty \gamma^{p-1} \frac{U\Big(m_1 L, m_1 L + 1 + m, \mathcal{B}_0(\gamma)\Big)}{e^{\{q+\beta(n_1+1)\}\gamma}} \mathrm{G}_{r+y+1,n+1}^{n,y+1} \left[\mathcal{D}_7 (\gamma)^y \Big| \begin{smallmatrix} 1,\tau_3,\tau_5 \\ \tau_4,0 \end{smallmatrix} \right] d\gamma}_{\mathcal{J}_6} \tag{4.48}$$

$$P_{b33} = \mathcal{D}_6 \sum_{m=0}^{n_2+m_1-1} \sum_{i=0}^{M-1} x_i \underbrace{\int_0^\infty \gamma^{m+p-1} U\Big(m_1 L, m_1 L+1+m, \mathcal{B}_0(\gamma)\Big) e^{-\{q+\beta(n_1+1)\}\gamma} \mathcal{H}\left[\Omega_{I2}, \left(\frac{\gamma}{\mu_r}\right)^y \right] d\gamma}_{\mathcal{J}_7} \tag{4.49}$$

With some mathematical manipulations, $\int_0^\infty \gamma^{p-1} e^{-q\gamma} d\gamma$ can be evaluated as $\frac{\Gamma(p)}{q^p}$. To solve $\mathcal{J}_3$, the confluent hypergeometric function can be represented using the relation $U(a, a+n+1, \lambda) = \lambda^{-a} \sum_{z=0}^{n} \frac{(a)_z}{\lambda^z}$. Then on applying [30, Eq. (383.5)], a closed-form solution of $\mathcal{J}_3$ can be obtained. On the other hand, $\mathcal{J}_4$ can be solved with the help of [31, Eq. (07.34.21.0088.01)]. Furthermore, to obtain the closed-form expression of $\mathcal{J}_5$, the bivariate Fox-H function may be represented using [32, Eq. (1.1)], and the identities in [30, Eq. (3.381.4)] and [33, Eq (28–30)] can be applied. The integral $\mathcal{J}_6$ can be represented as product of three Meijer-G functions that can be further evaluated using [32, Eq. (2.3)]. Moreover, we can solve the integral $\mathcal{J}_7$ by applying [32, Eq. (1.1)], [30, Eq. (3.381.4)], and [33, Eq (28–30)].

After performing the mathematical manipulations, the closed-form expression for the BER is written as provided in (4.50)

$$
\begin{aligned}
\bar{P}_b = & \frac{K}{\Gamma m_1} \sum_{n_1=0}^{K-1} \sum_{n_2=0}^{n_1(m_1-1)} A_3 \Bigg[\frac{1}{2} - \frac{q^p}{2\Gamma(p)} \sum_{m=0}^{n_2+m_1-1} \\
& \times \sum_{z=0}^{m} A_5 \Bigg\{ \Gamma(m+p)[(\Omega_{I_1}/m_1)\beta(n_1+1)]^{-m-p} \\
& \times U\Big(m+p, m+p+1-z-m_1, \mathcal{B}_1 \Big) \\
& + \frac{[q+\{\beta(n_1+1)\}]^{-m-p}(\mathcal{D}_6)_z}{\Gamma(z+m_1 L)} \sum_{i=0}^{M-1} x_i \frac{(y\Omega_{I_2})^{i+\frac{1}{2}}}{(2\pi)^{\frac{y-1}{2}}} \\
& \times \Bigg(\mathrm{H}_{1,0:1,1:y+r+1,n+1}^{0,1:1,1:n,y+1} \left[\begin{array}{c|c|c|c} \tau_{14} & \tau_{16} & \tau_{15} & \frac{1}{\mathcal{B}_1}, \mathcal{D}_{13} \\ \text{—} & (0,1) & (\tau_4, [1]_{\tau_4}), (0,1) & \end{array} \right] \\
& - \mathrm{H}_{2,1:0,1:r+1,n+1:1,1}^{0,2:1,0:n,1:1,1} \left[\begin{array}{c|c|c|c|c} \tau_{12} & \text{—} & (1,1)(\tau_3, [1]_{\tau_3}) & \tau_{16} & \frac{1}{\Omega_{I2}}, \\ \tau_{13} & (0,1) & (\tau_3, [1]_{\tau_3})(0,1) & (0,1) & \end{array} \right. \\
& \times \left. \frac{\mathcal{D}_5}{\mu_r^y [q+\beta(n_1+1)]^y}, \mathcal{D}_8 \right] \Bigg) \Bigg\} - \frac{\mathcal{D}_6}{2\Gamma(p)} \sum_{i=0}^{M-1} x_i \\
& \times \Bigg\{ \frac{y^{i+p} {\Omega_{I2}}^{i+1}}{(2\pi)^{y-1}} \mathrm{G}_{2y+r+1,u+1}^{n,2y+1} \left[\mathcal{D}_9 \middle| \begin{smallmatrix} 1,\tau_8 \\ \tau_4,0 \end{smallmatrix} \right] \\
& - \mathrm{H}_{1,2:0,1:r+2,n}^{0,1:1,0:n,2} \left[\begin{array}{c|c|c|c} \tau_{11} & \text{—} & \tau_9 & \frac{1}{\Omega_{I2}}, \frac{\mathcal{D}_5}{(q\mu_r)^y} \\ \tau_{10} & (0,1) & (\tau_4, [1]_{\tau_4}) & \end{array} \right] \Bigg\} \Bigg] + \frac{\mathcal{D}_6}{2\Gamma(p)} \\
& \times \sum_{i=0}^{M-1} x_i \Bigg\{ \frac{y^{i+p} {\Omega_{I2}}^{i+1}}{(2\pi)^{y-1}} \mathrm{G}_{2y+r+1,u+1}^{n,2y+1} \left[\mathcal{D}_9 \middle| \begin{smallmatrix} 1,\tau_8 \\ \tau_4,0 \end{smallmatrix} \right] \\
& - \mathrm{H}_{1,2:0,1:r+2,n}^{0,1:1,0:n,2} \left[\begin{array}{c|c|c|c} \tau_{11} & \text{—} & \tau_9 & \frac{1}{\Omega_{I2}}, \frac{\mathcal{D}_5}{(q\mu_r)^y} \\ \tau_{10} & (0,1) & (\tau_4, [1]_{\tau_4}) & \end{array} \right] \Bigg\}
\end{aligned}
\tag{4.50}
$$

In (4.50), $\mathrm{H}^{0,m_1:m_2,n_2:m_3,n_3:m_4,n_4}_{p_1,q_1:p_2,q_2:p_3,q_3:p_4,q_4}$ [.] is the trivariate Fox's-H function given in [33, Eq. (28)] and $(a)_z = \frac{\Gamma(a+z)}{\Gamma(a)}$. Also, various constants and parameters involved are defined as $A_5 = \binom{m}{z}\frac{(m_1L)_z\{\beta(n_1+1)\}^m}{m!}\left(\frac{m_1}{\Omega_{I1}}\right)^{-z}$, $\mathcal{B}_1 = \frac{q+\beta(n_1+1)}{(\Omega_{I1}/m_1)\beta(n_1+1)}$, $\mathcal{D}_9 = \mathcal{D}_7(\frac{y}{q})^y$, $\tau_8 = \left[\tau_3, \tau_5, \frac{1-p}{y}, \ldots, \frac{y-p}{y}\right]$, $\mathcal{D}_{13} = \frac{\mathcal{D}_7}{[q+\beta(n_1+1)]^y}$ $\tau_9 = [(1,1)(1-p,y)(\tau_3,[1]_{\tau_3})]$, $\tau_{10} = [(-i-1:1,y)]$, $\tau_{11} = [(-i:1,y)]$, $\tau_{12} = [(-i:1,y,0)(1-m-p:0,y,1)]$, $\tau_{13} = [(-i-1:1,y,0)]$, $\tau_{14} = (1-m-p:1,y)$, $\tau_{15} = (1,1)(\tau_3,[1]_{\tau_3})(\tau_5,[1]_{\tau_5})$, $\tau_{16} = (1-z-m_1L,1)$. As per the derived expression, the error rate is impacted by both the sources of interference in terms of INR and numbers. The BER depends on the atmospheric turbulence and pointing error on the FSO link modeled using the D-GG statistical model. This also means that the derived results can be modified by altering the values of the D-GG model parameters to gain insights into the system performance over Gamma-Gamma, log-normal, K, negative exponential, and Rician fading models. The BER expression also accounts for the number of RF users that simultaneously communicate with the relay node.

4.4.5 Asymptotic Bit Error Rate

To get more insights about the BER results, a simpler form can be derived by considering high SNR approximations. From the asymptotic expansion of function $U(m+p, m+p+1-z-m_1L, \mathcal{B}_1)$, the most significant term comes when $z=0$. Therefore, we have $U(m+p, m+p+1-z-m_1L, \mathcal{B}_1) \simeq \frac{\Gamma(m+p-m_1L)}{\Gamma(m+p)}\mathcal{B}_1{}^{-m-p+m_1L}$. Next, an approximation of bivariate Fox's H function may be developed considering the fact that the arguments become arbitrarily small as $\bar{\gamma}_k, \mu_r \to \infty$. Initially, with the help of [34], the bivariate Fox's H function can be represented in complex integral form as given in (4.51).

After some mathematical manipulations the Fox's H function involved in (4.52) can be expressed in terms of hypergeometric function as $\mathrm{H}^{1,2}_{2,1}\left[\frac{1}{\mathcal{B}_1}\middle| \begin{matrix}(1-m-p+ys_2,1)(1-z-m_1L,1)\\(0,1)\end{matrix}\right] = \frac{\Gamma(m+p-ys_2)\Gamma(z+m_1L)}{(2\pi i)} {}_2F_0(m+p-ys_2, z+m_1L;\ ;-\frac{1}{\mathcal{B}_1})$. Considering $\mathbf{s_2}$ to be the dominant pole from τ_4, the high SNR approximation can be obtained as stated in (42).

$$\mathrm{H}^{0,1:1,1:n,y+1}_{1,0:1,1:y+r+1,n+1}\left[\begin{matrix}\tau_{14}\\ \text{---}\end{matrix}\middle|\begin{matrix}\tau_{16}\\(0,1)\end{matrix}\middle|\begin{matrix}\tau_{15}\\(\tau_4,[1]_{\tau_4}),(0,1)\end{matrix}\middle|\frac{1}{\mathcal{B}_1},\mathcal{D}_{13}\right]$$
$$= \frac{1}{(2\pi i)^2}\int_{\mathcal{L}_\infty}\int_{\mathcal{L}_\in}\Gamma(m+p-s_1-ys_2)\Gamma(z-m_1L-s_1)$$
$$\times\frac{\Gamma(s_1)\prod_{j=1}^{n}\Gamma(\tau_{4,j}+s_2)\prod_{j=1}^{y}\Gamma(\tau_{5,j}-s_2)}{(s_2)\prod_{j=1}^{y}\Gamma(\tau_{3,j}+s_2)}\left(\frac{1}{\mathcal{B}_1}\right)^{-s_1}(\mathcal{D}_{13})^{-s_2}\,ds_1ds_2 \tag{4.51}$$

$$= \mathrm{H}_{2,1}^{1,2}\left[\frac{1}{\mathcal{B}_1}\middle| \begin{smallmatrix}(1-m-p+ys_2,1)(1-z-m_1L,1)\\(0,1)\end{smallmatrix}\right] \frac{1}{(2\pi i)}$$

$$\int_{\mathcal{L}_\in} \frac{\prod_{j=1}^{n}\Gamma(\tau_{4,j}+s_2)\prod_{j=1}^{y}\Gamma(\tau_{5,j}-s_2)}{(s_2)\prod_{j=1}^{y}\Gamma(\tau_{3,j}+s_2)}(\mathcal{D}_{13})^{-s_2}\,ds_2 \tag{4.52}$$

$$\underset{\bar{\gamma}_k,\mu_r\to\infty}{\simeq} \frac{\Gamma(m+p-y\mathbf{s_2})\Gamma(z+m_1L)}{(2\pi i)}{}_2F_0\left(m+p-y\mathbf{s_2}, z+m_1L;\quad ;-\frac{1}{\mathcal{B}_1}\right)$$
$$\times\frac{\prod_{j=1}^{n}\Gamma(\tau_{4,j}+\mathbf{s_2})\prod_{j=1}^{y}\Gamma(\tau_{5,j}-\mathbf{s_2})}{\mathbf{s_2}\prod_{j=1}^{y}\Gamma(\tau_{3,j}+\mathbf{s_2})}$$
$$\times(\mathcal{D}_{13})^{\mathbf{s_2}} \tag{4.53}$$

Utilizing the asymptotic expression of bivariate Fox's H function $\mathcal{H}\left[\Omega_{I2},\left(\frac{\gamma}{\mu_r}\right)^{y}\right]$ given in $\mathcal{J}_2$ in (4.35), the trivariate Fox's H function can be further approximated. Accordingly, the integral generating trivariate Fox's H function can be simplified for high SNR regime as expressed in (4.53). Moreover, the trivariate Fox-H function involved in BER results can be approximated as stated in (4.54).

$$\mathrm{H}_{2,1:0,1:r+1,n+1:1,1}^{0,2:1,0:n,1:1,1}\left[\begin{matrix}\tau_{12}\\\tau_{13}\end{matrix}\middle|\begin{matrix}\text{—}\\(0,1)\end{matrix}\middle|\begin{matrix}(1,1)(\tau_3,[1]_{\tau_3})\\(\tau_3,[1]_{\tau_3})(0,1)\end{matrix}\middle|\begin{matrix}\tau_{16}\\(0,1)\end{matrix}\middle|\frac{1}{\Omega_{I2}},\frac{\mathcal{D}_5}{\mu_r^{y}[q+\beta(n_1+1)]^{y}},\right.$$
$$\left.\times\mathcal{D}_8\right]\underset{\bar{\gamma}_k,\mu_r\to\infty}{\simeq}\frac{\Gamma(1+i+y\mathbf{s})}{\Gamma(2+i+y\mathbf{s})}\left(\frac{\mathcal{D}_5}{\mu_r^{y}}\right)^{+\mathbf{s}}$$
$$\times\frac{\prod_{k=1}^{n}\Gamma(\tau_{4,k}-\mathbf{s})\Gamma(\mathbf{s})}{\prod_{k=2}^{r+2}\Gamma(\tau_{3,k}-\mathbf{s})\Gamma(1+\mathbf{s})}{}_1F_1\left(1+i+y\mathbf{s},2+i+y\mathbf{s}:-\frac{1}{\Omega_{I2}}\right)$$
$$\times\int_0^\infty \gamma^{m+p-y\mathbf{s}-1}\frac{\mathrm{G}_{1,1}^{1,1}\left[\frac{\Omega_{I1}}{m_1}\beta(n_1+1)\gamma\middle|\begin{smallmatrix}1-z-m_1L\\0\end{smallmatrix}\right]}{\exp\left(-\{q+\beta(n_1+1)\gamma\}\right)}d\gamma \tag{4.54}$$

Utilizing these steps, the asymptotic expression for the BER can be formulated. This expression is only in terms of elementary functions, and simulation of the asymptotic BER expression is less time-consuming at high SNR values as compared to the exact expression.

4.4.6 *Ergodic Capacity Analysis*

The maximum error-free transmission rate is defined as the capacity of the system. Ergodic capacity of a wireless communication system is defined as $\mathbb{E}(\log_2(1+\gamma))$. Furthermore, it can be expressed in terms of PDF as [37]

$$\bar{\mathcal{C}} = \int_0^\infty \log_2(1+\delta\gamma) f_{\gamma_{e2e}}(\gamma) \tag{4.55}$$

where $\delta = 1$ or $\frac{e}{e\pi}$ given the heterodyne or IM/DD detection scheme is used at the receiver [38]. Since analysis using exact function of $f_{\gamma_{e2e}}(\gamma)$ is not tractable, so the end-to-end PDF is approximated as per following discussion. On replacing $f_{e_2e}(\gamma)$ in (4.55), the ergodic capacity can be given as $\bar{\mathcal{C}} = \bar{\mathcal{C}_1}+\bar{\mathcal{C}_2}-\bar{\mathcal{C}_3}-\bar{\mathcal{C}_4}$. On substituting the overall PDF $f_{e_2e}(\gamma)$ in (4.55), we get the integrals $\bar{\mathcal{C}_1}$, $\bar{\mathcal{C}_2}$, $\bar{\mathcal{C}_3}$, $\bar{\mathcal{C}_4}$.

$$\bar{\mathcal{C}_1} = \sum_{n_1=0}^{K-1} \sum_{n_2=0}^{n_1(m_1-1)} A_2 \underbrace{\int_0^\infty \gamma^{m_1+n_2-1} e^{[-\beta(n_1+1)\gamma]} U\big(m_1 L, m_1 L+1+m_1+n_2, \mathcal{B}_0(\gamma)\big) \log_2(1+\delta\gamma) d\gamma}_{\mathcal{J}_8} \tag{4.56}$$

To evaluate $\mathcal{J}_8$ in (4.56), we express $U\big(m_1 L, m_1 L + 1 + m_1 + n_2, \mathcal{B}_0(\gamma)\big) = \mathcal{B}_0(\gamma)^{-m_1 L} \sum_{z=0}^{n_2+m_1} \binom{n_2+m_1}{z} \frac{(m_1 L)_z}{(\mathcal{B}_0(\gamma))^{z+m_1 L}}$. The argument of confluent hypergeometric function $\mathcal{B}_0(\gamma)$ can be represented in terms of Meijer-G function as

$$\left(\beta(n_1+1)\gamma + \frac{m_1}{\Omega_{I1}}\right)^{-z-m_1 L} = \frac{\left(m_1/\Omega_{I_1}\right)^{-z-m_1 L}}{\Gamma(z+m_1 L)} \mathrm{G}_{1,1}^{1,1}\left[\frac{\Omega_{I1}}{m_1}\beta(n_1+1)\gamma \middle| \begin{matrix} 1-z-m_1 L \\ 0 \end{matrix}\right] \tag{4.57}$$

Using [31, Eq. (07.34.03.0456.01)], to express $\log_2(1+\delta) = \mathrm{G}_{2,2}^{1,2}\left[\delta\gamma \middle| \begin{matrix} 1,1 \\ 1,0 \end{matrix}\right]$, and substituting the Meijer-G representation of exponential function $e^{[-\beta(n_1+1)\gamma]} = \mathrm{G}_{0,1}^{1,0}\left[\beta(n_1+1)\gamma \middle| \begin{matrix} - \\ 0 \end{matrix}\right]$ as per [31, Eq. (07.34.03.0228.01)], we can formulate the solution of $\mathcal{J}_8$ with the help of [32, Eq. (2.3)]. We now move toward calculating $\bar{\mathcal{C}_2} = \bar{\mathcal{C}_{21}} - \bar{\mathcal{C}_{22}}$. The integral $\bar{\mathcal{C}_{21}}$ can be defined as

$$\bar{\mathcal{C}_{21}} = \mathcal{D}_{11} \sum_{i=0}^{M-1} x_i \mathcal{D}_{12} \mathcal{J}_9 \tag{4.58}$$

where

$$\mathcal{J}_9 = \int_0^\infty \mathrm{G}_{r+y, r(\lambda+\sigma+1)}^{r(\lambda+\sigma+1), y}\left[\mathcal{D}_8 \gamma^y \middle| \begin{matrix} \tau_6 \\ \tau_7 \end{matrix}\right] \log_2(1+\delta\gamma) d\gamma$$

In order to obtain the closed-form solution of $\mathcal{J}_9$ to evaluate $\bar{C}_{21}$, we substitute the Meijer-G form of $\log_2(1+\delta)$ and then apply [31, Eq. (07.34.21.0013.01)]. The description of $\bar{C}_{22}$ is given below

$$\bar{C}_{22} = \mathcal{D}_{11}\sum_{i=0}^{M-1} x_i \underbrace{\int_0^\infty \mathcal{H}\left[\Omega_{I2}, \mathcal{D}_2 z^y \left(\frac{\gamma}{\mu_r}\right)^{y/r}\right]\log_2(1+\delta\gamma)d\gamma}_{\mathcal{J}_{10}} \tag{4.59}$$

With the knowledge of [34, Eq. (2.56)], the bivariate Fox-H function in $\mathcal{J}_{10}$ can be expressed in two-fold Mellin-Barnes representation of the H-Fox function. Similarly, $\mathrm{G}_{2,2}^{1,2}\left[\delta\gamma\middle|\begin{smallmatrix}1,1\\1,0\end{smallmatrix}\right]$ can be represented in Fox-H function using [31, Eq. (07.34.26.0008.01)] and denoted in contour integral using [34, Eq. (1.2–1.3)]. After some manipulations, [32, Eq. (2.3)] can be applied to obtain closed-form expression for $\bar{C}_{22}$. It is defined that $\bar{C}_3 = \bar{C}_{31} - \bar{C}_{32}$ since $\bar{C}_{31} = \mathcal{D}_6\sum_{n_1=0}^{K-1}\sum_{n_2=0}^{n_1(m_1-1)} A_2\sum_{i=0}^{M-1} x_i\mathcal{D}_{10}\mathcal{J}_{11}$, where $\mathcal{J}_{11}$ is defined in (4.60). After using (4.57), the integrand $\mathcal{J}_{11}$ can be concluded to involve product of the three independent Meijer G functions that can be solved with the aid of [33].

Next, an approximate solution of $\mathcal{J}_{12}$ in (4.67) can be formulated on substituting approximation of $\mathcal{H}\left[\Omega_{I2}, \left(\frac{\gamma}{\mu_r}\right)^y\right]$ from (4.35). After applying [30, Eq. (3.381.4)] and using [33, Eq (28–30)], the integral $\mathcal{J}_{12}$ can be derived in terms of the bivariate Fox-H function.

$$\begin{aligned}\mathcal{J}_{11} = &\int_0^\infty \gamma^{m_1+n_2-1} e^{-[\beta(n_1+1)\gamma]} U\big(m_1 L, m_1 L + 1 + m_1 + n_2, \mathcal{B}_0(\gamma)\big)\\ &\times \log_2(1+\delta\gamma)\mathrm{G}_{r+y+1,n+1}^{n,y+1}\left[\mathcal{D}_7(\gamma)^y\middle|\begin{smallmatrix}1,\tau_3,\tau_5\\ \tau_4,0\end{smallmatrix}\right]d\gamma\end{aligned} \tag{4.60}$$

Further to this, after multiplying $f_{\gamma_2}^o(\gamma)$ with the CDF on the RF link $F_{\gamma_1}^o(\gamma)$, we can formulate $\bar{C}_4$ as

$$\bar{C}_4 = \frac{K}{\Gamma m_1}\sum_{n_1=0}^{K-1}\sum_{n_2=0}^{n_1(m_1-1)} A_3\bar{C}_2 + \bar{C}_{41} - \bar{C}_{42} \tag{4.61}$$

where $\bar{C}_{41}$ and $\bar{C}_{42}$ can be defined as

$$\bar{C}_{41} = \mathcal{D}_{11}\frac{K}{\Gamma m_1}\sum_{n_1=0}^{K-1}\sum_{n_2=0}^{n_1(m_1-1)} A_3 \sum_{m=0}^{n_2+m_1-1}\frac{\{(n_1+1)\beta\}^m}{m!}\sum_{i=0}^{M-1} x_i\mathcal{D}_{12}\mathcal{J}_{13} \tag{4.62}$$

$$\bar{C}_{42} = \mathcal{D}_{11}\frac{K}{\Gamma m_1}\sum_{n_1=0}^{K-1}\sum_{n_2=0}^{n_1(m_1-1)} A_3 \sum_{m=0}^{n_2+m_1-1}\frac{\{(n_1+1)\beta\}^m}{m!}\sum_{i=0}^{M-1} x_i\mathcal{J}_{14} \tag{4.63}$$

where integrals $\mathcal{J}_{13}$ and $\mathcal{J}_{14}$ are shown in (4.64) and (4.65), respectively.

$$\mathcal{J}_{13} = \int_0^\infty \gamma^m e^{(-\beta(n_1+1)\gamma)} U\left(m_1 L, m_1 L + 1 + m, \mathcal{B}_0(\gamma)\right) \mathrm{G}_{r+y, r(\lambda+\sigma+1)}^{r(\lambda+\sigma+1), y} \times \left[\mathcal{D}_8 \gamma^y \middle| \begin{smallmatrix} \tau_6 \\ \tau_7 \end{smallmatrix}\right] d\gamma \tag{4.64}$$

$$\mathcal{J}_{14} = \int_0^\infty \gamma^m e^{(-\beta(n_1+1)\gamma)} U\left(m_1 L, m_1 L + 1 + m, \mathcal{B}_0(\gamma)\right) \times \mathcal{H}\left[\Omega_{I2}, \mathcal{D}_2 z^y \left(\frac{\gamma}{\mu_r}\right)^{y/r}\right] d\gamma \tag{4.65}$$

With the aid of (4.57) and [32, Eq. (2.3)], a closed-form solution to $\mathcal{J}_{13}$ can be established. An approximate analysis of $\mathcal{J}_{14}$ can be performed by representing the bivariate Fox-H function $\mathcal{H}\left[\Omega_{I2}, \mathcal{D}_2 z^y \left(\frac{\gamma}{\mu_r}\right)^{y/r}\right]$ in terms of its dominant pole. More specifically, assuming that $\mathbf{s}_3$ is $\min(\tau_1)$, the bivariate Fox's H function involved in the PDF expression of $f_{\gamma_2}^{o}(\gamma)$ can be approximated as given in (4.66).

$$\mathcal{H}\left[\Omega_{I2}, \mathcal{D}_2 z^y \left(\frac{\gamma}{\mu_r}\right)^{y/r}\right] \simeq \frac{\prod_{j=1}^{\lambda+\sigma+1} \Gamma\left(\tau_{1,j} + \mathbf{s}_3\right)}{\Gamma\left(\tau_{2,j} + \mathbf{s}_3\right)} 1 \times F_1\left(i + 2 - (y/r)\mathbf{s}_3; i + 3 - (y/r)\mathbf{s}_3; -\frac{1}{\Omega_{I_2}}\right) \left(\mathcal{D}_2 z^y \left(\frac{\gamma}{\mu_r}\right)^{y/r}\right)^{-\mathbf{s}_3} \tag{4.66}$$

This approximation can be substituted in (4.65), and then by applying [31, Eq. (07.34.21.0013.01)], an approximate solution of $\mathcal{J}_{14}$ can be obtained.

$$\bar{C}_{32} = \mathcal{D}_6 \sum_{n_1=0}^{K-1} \sum_{n_2=0}^{n_1(m_1-1)} A_2 \sum_{i=0}^{M-1} x_i \underbrace{\left(\int_0^\infty \gamma^{m_1+n_2-1} \frac{U\left(m_1 L, m_1 L + 1 + m_1 + n_2, \mathcal{B}_0(\gamma)\right)}{e^{[\beta(n_1+1)\gamma]}} \log_2(1+\delta\gamma) \mathcal{H}\left[\Omega_{I2}, \left(\frac{\gamma}{\mu_r}\right)^y\right] d\gamma\right)}_{\mathcal{J}_{12}} \tag{4.67}$$

After plugging various terms in the definition of ergodic capacity, the closed-form expression can be written in (4.68) as

$$
\begin{aligned}
\bar{\mathcal{C}} = & \sum_{n_1=0}^{K-1}\sum_{n_2=0}^{n_1(m_1-1)} A_2\beta^{n_2}\sum_{z=0}^{n_2+m_1}\binom{n_2+m_1}{z}\frac{(m_1L)_z\{\beta(n_1+1)\}^{-n_2-m_1}}{\Gamma(z+m_1L)} \\
& \times\left[\mathrm{H}_{\mathbf{h}_1}^{\mathbf{h}_2}\left[\begin{array}{c|c|c|c} \tau_{17} & (1,1)(1,1) & \tau_{17} & \frac{\delta}{\beta(n_1+1)},\frac{\Omega_{I_1}}{m_1} \\ \text{—} & (1,1)(0,1) & (0,1) & \end{array}\right]\right. \\
& -\mathcal{D}_6\sum_{i=0}^{M-1}x_i\left\{\mathcal{D}_{10}\mathrm{H}_{\mathbf{h}_3}^{\mathbf{h}_4}\left[\begin{array}{c|c|c|c|c} \tau_{18} & \tau_{16} & (1,1)(1,1) & (1,1)(\tau_3,[1]_{\tau_3}) & \frac{\Omega_{I1}}{m_1}, \\ \text{—} & (0,1) & (1,1)(0,1) & (\tau_4,[1]_{\tau_4})(0,1) & \end{array}\right.\right. \\
& \left.\times\frac{\delta}{[\beta(n_1+1)]},\frac{\mathcal{D}_7}{[\beta(n_1+1)]^y}\right] \\
& +\{\beta(n_1+1)\}^{y\mathbf{s}}\left(\frac{\mathcal{D}_5}{\mu_r^y}\right)^{\mathbf{s}}\frac{\Gamma(1+i+y\mathbf{s})}{\Gamma(2+i+y\mathbf{s})}\Lambda_3\,{}_1F_1 \\
& \left(1+i+y\mathbf{s},2+i+y\mathbf{s}:-\frac{1}{\Omega_{I2}}\right) \\
& \left.\left.\times\mathrm{H}_{\mathbf{h}_5}^{\mathbf{h}_6}\left[\begin{array}{c|c|c|c} \tau_{19} & \tau_{17} & (1,1)(1,1) & \frac{\Omega_{I_1}}{m_1},\frac{\delta}{\beta(n_1+1)} \\ \text{—} & (0,1) & (1,1)(0,1) & \end{array}\right]\right\}\right] \\
& +\mathcal{D}_{11}\sum_{i=0}^{M-1}x_i\left[\mathcal{D}_{12}\frac{y^{-\frac{1}{2}}\delta^{-1}}{(2\pi)^{y-1}}\mathrm{G}_{r+2y,r(\lambda+\sigma+1)+2v}^{r(\lambda+\sigma+1)+2v,2y}\left[\frac{\mathcal{D}_8}{\delta^{-1}}\middle|\begin{array}{c}\tau_8\\ \tau_{20}\end{array}\right]-\delta^{-1}\right. \\
& \left.\times\mathrm{H}_{\mathbf{h}_7}^{\mathbf{h}_8}\left[\begin{array}{c|c|c|c} \tau_{21} & \text{—} & \tau_{23} & \frac{1}{\Omega_{I_1}},\frac{\mathcal{D}_2z^y}{(\delta\mu_r)^{\frac{v}{r}}} \\ \tau_{22} & (0,1) & \tau_{24} & \end{array}\right]\right]+\frac{K}{\Gamma m_1}\sum_{n_1=0}^{K-1}\sum_{n_2=0}^{n_1(m_1-1)}A_3 \\
& \times\sum_{m=0}^{n_2+m_1-1}\sum_{z=0}^{m}A_5\frac{\mathcal{D}_{11}}{\Gamma(z+m_1L)}\sum_{i=0}^{M-1}x_i\left[\frac{\mathcal{D}_{12}}{\{\beta(n_1+1)\}}\right. \\
& \times\mathrm{H}_{\mathbf{h}_9}^{\mathbf{h}_{10}}\left[\begin{array}{c|c|c|c|c} \tau_{25} & \tau_{16} & (1,1)(1,1) & (\tau_6,[1]_{\tau_6}) & \frac{\Omega_{I1}}{m_1},\frac{\delta}{[\beta(n_1+1)]},\frac{\mathcal{D}_8}{[\beta(n_1+1)]^y} \\ \text{—} & (0,1) & (1,1)(0,1) & (\tau_7,[1]_{\tau_7}) & \end{array}\right] \\
& -\frac{\Gamma(i+2+\frac{v}{r}\mathbf{s})}{\Gamma(i+3+\frac{v}{r}\mathbf{s})}\left(\frac{\mathcal{D}_2z^y}{\mu_r^{\frac{v}{r}}}\right)^{\mathbf{s}} \\
& \times\frac{\prod_{j=1}^{\lambda+\sigma+1}\Gamma(\tau_{1,j}-\mathbf{s})}{\Gamma(\tau_2-\mathbf{s})}{}_1F_1\left(i+2+\frac{v}{r}\mathbf{s};i+3+\frac{v}{r}\mathbf{s};\frac{-1}{\Omega_{I_2}}\right) \\
& \left.\times\mathrm{H}_{\mathbf{h}_{11}}^{\mathbf{h}_{12}}\left[\begin{array}{c|c|c|c} \tau_{26} & \tau_{16} & (1,1)(1,1) & \frac{\Omega_{I_1}}{m_1},\frac{\delta}{\beta(n_1+1)} \\ \text{—} & (0,1) & (1,1)(0,1) & \end{array}\right]\right]
\end{aligned}
\tag{4.68}
$$

In (4.68), $\tau_{17} = (1 - n_2 - m_1 : 1, 1)$, $\tau_{18} = (1 - n_2 - m_1 : 1, 1, v)$, $\tau_{19} = (1 - n_2 - m_1 + \mathbf{s} : 1, 1)$, $\tau_{20} = [\tau_7, \frac{-1}{y}, \ldots, \frac{y-1}{y}, \frac{-1}{y}, \ldots, \frac{y-1}{y}]$, $\tau_{21} = (-i - 1 : 1, v/r)$, $\tau_{22} = (-i - 2 : 1, v/r)$, $\tau_{23} = (-1, v/r)(\tau_2, [1]_{\tau_2})(0, v/r)$, $\tau_{24} = (\tau_1, [1]_{\tau_1})(-1, v/r)(-1, v/r)$, $\tau_{25} = (-m : 1, 1, v)$,$\tau_{26} = (-m - \frac{v\mathbf{s}}{r} : 1, 1)$, $\{\mathbf{h}_1, \mathbf{h}_2\} = \{(1, 0 : 2, 2 : 1, 1), (0, 1 : 1, 2 : 1, 1)\}$, $\{\mathbf{h}_3, \mathbf{h}_4\} = \{(1, 0 : 1, 1 : 2, 2 : \mathrm{y+1+r}, \mathrm{n+1}), (0, 1 : 1, 1 : 1, 2 : \mathrm{n}, \mathrm{y+1})\}$, $\{\mathbf{h}_5, \mathbf{h}_6\} = \{(1, 0 : 1, 1 : 2, 2), (0, 1 : 1, 1 : 1, 2)\}$, $\{\mathbf{h}_7, \mathbf{h}_8\} = \{(1, 1 : 1, 0 : \lambda + \sigma + 3, 1), (1, 1 : 0, 1 : 3, \lambda + \sigma + 3)\}$, $\{\mathbf{h}_9, \mathbf{h}_{10}\} = \{(1, 0 : 1, 1 : 2, 2 : r + y, r(\lambda + \sigma + 1)), (0, 1 : 1, 1 : 1, 2 : r(\lambda + \sigma + 1), y)\}$, $\{\mathbf{h}_{11}, \mathbf{h}_{12}\} = \{(1, 0 : 1, 1 : 2, 2), (0, 1 : 1, 1 : 1, 2)\}$. The ergodic capacity depends on the number of RF users, the number of interferers involved, RF fading, and turbulence on optical link along with the pointing error. The nature of the derived results suggests that increase in strength of interference reduces the channel capacity of the system. The derived results can be used to also validate the performance of the system when the RF link is subjected to Rayleigh fading as Nakagami-m model has Rayleigh distribution as special case. Similarly, the FSO link model used for the derivation of capacity expression is D-GG model that is also generalized in nature and can represent the atmospheric turbulence for statistical models like negative exponential, Gamma-Gamma, log-normal among others.

4.5 Results

The scheme that has been analyzed in the present chapter models the FSO link using the D-GG turbulence model. The parameters selected for representing the FSO link are as follows: $\alpha_1 = 2.169$; $\alpha_2 = 1$; $\beta_1 = 0.55$; $\beta_2 = 2.35$; $\Omega_1 = 1.5793$; $\Omega_2 = 1$; $\lambda = 13$; $\sigma = 6$. The BER results have been plotted in Fig. 4.1. Here, it has been assumed that $K = 2$ RF users are multiplexed onto the FSO channel. The user

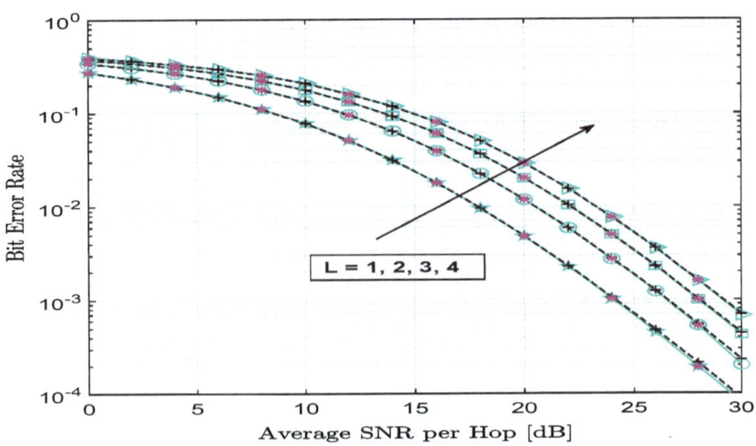

Fig. 4.1 Impact of increase in the number of interferers

diversity between the two radio users enhances the user experience by decreasing the BER. The INR for the interferer that is in the vicinity of the relay node has been taken as 10 dB. On the other hand, the INR of the CCI at the destination node has been assumed to be 1 dB. The objective of the characteristic plot is to analyze the impact of the increase in the number of interfering devices near the relay node. Therefore, the number of CCIs has been varied as $L = 1, 2, 3, 4$. It can be noticed from the plot that with every increase in interfering signal, the BER response changes and shows a greater value for each average SNR value. This clearly suggests the impact of interference on the error rate. The results are verified using Monte Carlo simulation, and the dotted curves represent the asymptotic analysis results. The asymptotic analysis results follow with the exact results for almost all the practical ranges of average SNR values. This suggests that the approximate expression for the BER that has been derived for the considered system model can be used efficiently with less computational complexity as compared to the exact results.

The impact of pointing error and demodulation scheme on optical link has been highlighted in Fig. 4.2. For the purpose of analysis, the number of CCIs at both the relay and destination nodes is assumed as $L = M = 2$. Also, the interfering signal strength with respect to noise is 2 dB each. The pointing error fading has been captured with two values of the ξ parameters. The small misalignment error has been modeled using $\xi = 1.02$, whereas $\xi = 0.54$ is used for representing large pointing error on the FSO link. Moreover, the constant r denotes the type of optical demodulation on the optical wireless link. For the considered model parameters, the magnitude of pointing error dominates the performance as compared to the type of optical demodulation scheme. When the pointing error parameter is $\xi = 1.0206$, the BER experienced by the user is less for both the types of optical demodulation schemes. In addition to this, the intensity based optical demodulation scheme has the worst effect when the pointing error parameter is $\xi = 0.5427$. The Monte Carlo simulation results are also depicted in the figure, which relate closely with the theoretical results demonstrated in the plot.

Another way of investigating the performance based on the aperture of the receiver is by analyzing the BER as a function of ω/a ratio. This has been presented in Fig. 4.3. Furthermore, both kinds of optical demodulation schemes have been compared in the plot. The BER curve decreases as the average SNR per hop is increased. Also, two combinations of the apertures sizes of the optical receiver have been considered as $\omega/a = [15, 5]$. Of course, the larger the ratio of ω/a, the reception of the optical signal is more reliable as the probability of the photons to be received by the photodetector is more. However, even if a larger ratio of the detector size is considered, the optical demodulation scheme that is employed places a crucial role. The ratio $\omega/a = 15$ with $r = 2$ shows worse BER performance when compared to $r = 1$ for the same detector aperture. Therefore, it is important for the system designer to opt for coherent demodulation process if the other requirements are met. Clearly, the worst performance is when the $\omega/a = 5$ and $r = 2$, as intensity demodulation can be opted with such a smaller detector aperture ratio. The high SNR approximate results have also been plotted on the same figure, and

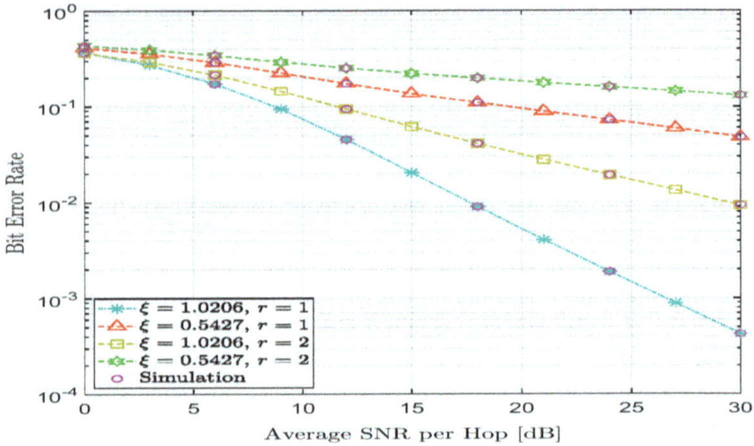

Fig. 4.2 Mixed RF/FSO AF relaying with D-GG optical channel with impact of pointing error and kind of optical demodulation scheme

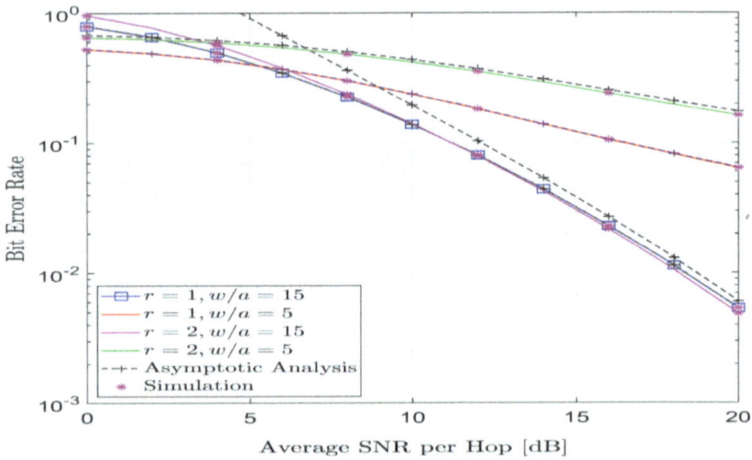

Fig. 4.3 Effect of variation in receiver aperture diameter along with type of optical modulation scheme

the increase in the average SNR axis makes the asymptotic results to approach the exact simulation values.

The ergodic capacity of the variable gain mixed RF/FSO AF relaying system has been shown in Fig. 4.4. The capacity relates to the throughput that can be established from the network. The derived approximation has been numerically evaluated and confirmed using the Monte Carlo simulation. The plot basically investigates the impact of type of optical demodulation (r), the number of interferers on the relay

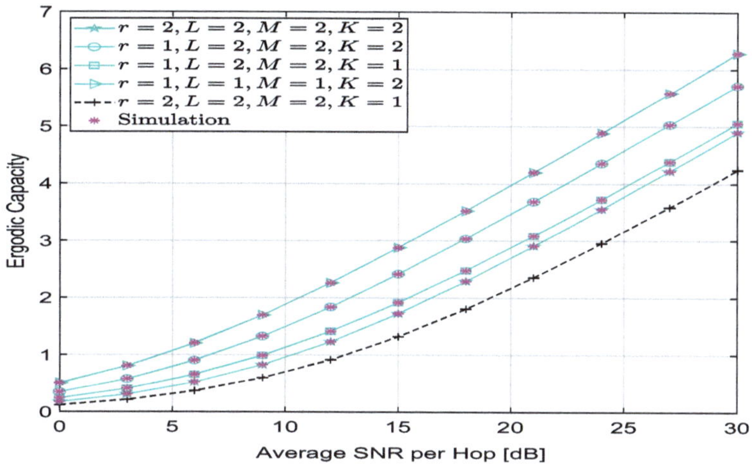

Fig. 4.4 Ergodic capacity of the variable gain mixed RF/FSO AF relaying system for varying set of parameters

(L), and the number of CCIs on the destination node (M) along with the number of RF users who transmit simultaneously in the user diversity scheme. There is substantial improvement in the capacity offered by the system when the number of RF users increases from $K = 1$ to $K = 2$. However, the major concern is the optical demodulation scheme employed. When the constant r is set to 2, which represents the IM/DD optical demodulation, the performance is much inferior to that when coherent demodulation scheme is used. Another important factor that governs the performance is the number of interferers available in the vicinity of the relay and destination nodes. When $L = 2$ and $M = 2$, the estimated capacity is much lower as compared to the case when $L = M = 1$. It can be deduced from the plot that when $L = 2$ and $M = 2$, and $K = 1$, the coherent demodulation does not provide a performance compared to when K is increased to 2. The best performance is clearly when coherent demodulation is considered on the FSO link and two RF users transmit the data with $L = M = 1$ interferers.

Apart from the number of interfering signals, the signal strength that works in opposition with the desired signal is also an important metric. Figure 4.5 demonstrates the capacity for varying strength of interfering signal Ω_{I1}. For demonstration purpose, the interference is assumed to be $\Omega_{I1} = [0, 3, 5]$ dB. As the strength of average INR of $L = 2$ interferers increases, a reduction in the ergodic capacity can be seen in the plot. For the analysis, it is considered that D-GG parameters are set as $\alpha_1 = 2.169; \alpha_2 = 1; \beta_1 = 0.55; \beta_2 = 2.35; \Omega_1 = 1.5793; \Omega_2 = 1; \lambda = 13; \sigma = 6$. Apart from the strength on interference, the user diversity can be utilized to boost the signal strength of the desired signal. This has been shown in the plot as a comparison between ergodic capacity values for $K = 1$ and $K = 2$ for the same level of interference $\Omega_{I1} = 5$ dB. Although the throughput of variable gain AF relaying system is less than the capacity values when

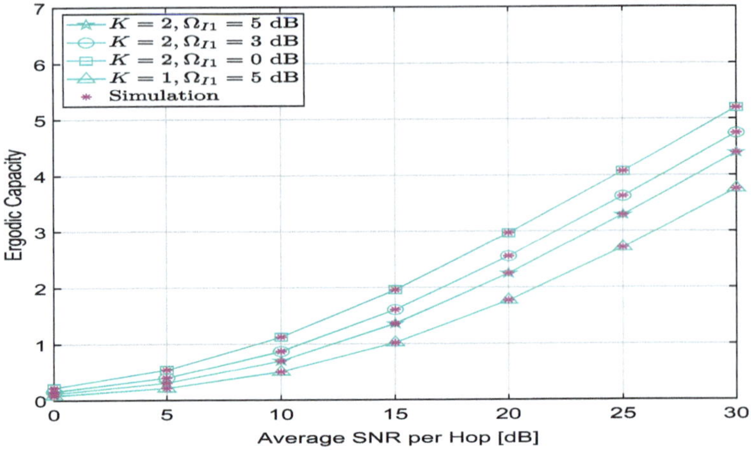

Fig. 4.5 Ergodic capacity of the mixed RF/FSO AF relaying system over D-GG channel for varying strength of interference

$\Omega_{I1} = [5]$ dB, an improvement in capacity values can be seen in the plot when the number of RF users is increased from $K = 1$ to $K = 2$. This means that although the interference cannot be reduced, the increase in the number of RF users has the capability to enhance the capacity of the system.

4.6 Conclusion

The present chapter analyzed mixed RF/FSO AF cooperative relaying systems when interference is present at relay and destination nodes. The assumption of interference on both the nodes makes the system more practical. Also, the RF side of the cooperative relay is generally scarce in bandwidth availability. Therefore, multiple RF users can be deployed simultaneously to be multiplexed onto a single FSO carrier. For this purpose, MUD scheme has been opted on the first hop of communication. The user diversity can enhance the reliability of the network by selecting the user with best channel condition. In addition to this, the optical link has been modeled using the D-GG channel. Since this channel model is generalized, the derived analytical results will serve as general results that can be used to find system performance for existing statistical models as special cases. For the considered system model, closed form of CDF has been presented in the existing literature but was been taken forward to analyze the error rate and throughput. In this chapter, a closed-form expression for the BER has been derived, which has been extended to find the asymptotic expression also. More importantly, the CDF expression, which yields the outage probability, has been modified to provide an asymptotic expression of the system, which has been considered to find the coding gain and diversity order for two separate cases. The chapter also stratifies an optimal power allocation scheme that can be deployed on the basis of the asymptotic outage probability results. The ergodic capacity analysis has also been carried out in the chapter for the mixed RF/FSO cooperative relaying system with dual level interference. The numerical results show that the system is heavily affected by the presence of interference.

References

1. A.A. Farid, S. Hranilovic, Outage capacity optimization for free-space optical links with pointing errors. J. Lightwave Technol. **25**, 1702–1710 (2007)
2. E. Lee, J. Park, D. Han, G. Yoon, Performance analysis of the asymmetric dual-hop relay transmission with mixed RF/FSO links. IEEE Photon. Technol. Lett. **23**(21), 1642–1644 (2011)
3. E. Zedini, I.S. Ansari, M.S. Alouini, Performance analysis of mixed nakagami-m and gamma-gamma dual-hop FSO transmission systems. IEEE Photonics J. **7**(1), 1–20 (2015)
4. Y.F. Al-Eryani, A.M. Salhab, S.A. Zummo, M.S. Alouini, Two-way multiuser mixed RF/FSO relaying: performance analysis and power allocation. IEEE/OSA J. Opt. Commun. Networking **10**(4), 396–408 (2018)
5. J. Abouei, K.N. Plataniotis, Multiuser diversity scheduling in free-space optical communications. J. Lightwave Technol. **30**, 1351–1358 (2012)
6. L. Chen, W. Wang, C. Zhang, Multiuser diversity over parallel and hybrid FSO/RF links and its performance analysis. IEEE Photonics J. **8**(3), 1–9 (2016)
7. L. Chen, W. Wang, Effective capacity of MIMO free-space optical systems over gamma gamma turbulence channels. Opt. Commun., Elsevier **382**, 450–454 (2017)
8. L. Yang, M.O. Hasna, I.S. Ansari, Unified performance analysis for multiuser mixed η—μ and $\mathcal{M}$—distribution dual-hop RF/FSO systems. IEEE Trans. Commun. **65**(8), 3601–3613 (2017)
9. A.H.A. El-Malek, A.M. Salhab, S.A. Zummo, M. Alouini, Effect of RF interference on the security-reliability trade-off analysis of multiuser mixed RF/FSO relay networks with power allocation. J. Lightwave Technol. **35**(9), 1490–1505 (2017)
10. L. Yang, X. Gao, M.S. Alouini, Performance analysis of free-space optical communication systems with multiuser diversity over atmospheric turbulence channels. IEEE Photonics J. **6**(2), 1–17 (2014)
11. A. Goel, R. Bhatia, On the performance of mixed user Diversity-RF/Spatial Diversity-FSO cooperative relaying AF systems. Opt. Commun. **477**, 126333 (2020)
12. A. Upadhya, V.K. Dwivedi, G. Singh, Multiuser diversity for mixed RF/FSO cooperative relaying in the presence of interference. Opt. Commun. **442**, 77–83 (2019)
13. A. Upadhya, V.K. Dwivedi, M.S. Alouini, Interference-limited mixed MUD-RF/FSO two-way cooperative networks over double generalized gamma turbulence channels. IEEE Commun. Lett. **442**, 1551–1555 (2019)
14. V.K. Tonk, A. Upadhya, P.K. Yadav, V.K. Dwivedi, Mixed mud-RF/FSO two way Dcode and forward relaying networks in the presence of co-channel interference. Opt. Commun. **464**, 125415 (2020)
15. A. Agustin, J. Vidal, Amplify-and-forward cooperation under interference-limited spatial reuse of the relay slot. IEEE Trans. Wirel. Commun. **7**(5), 1952–1962 (2008)
16. E. Soleimani-Nasab, M. Uysal, Generalized performance analysis of mixed RF/FSO cooperative systems. IEEE Trans. Wirel. Commun. **15**(1), 714–727 (2016)
17. M.I. Petkovic, A.M. Cvetkovic, G.T. Djordjevic, G.K. Karagiannidis, Outage performance of the mixed RF/FSO relaying channel in the presence of interference. Wirel. Pers. Commun. **96**(2), 2999–3014 (2017)

18. I. Trigui, N. Cherif, S. Affes, X. Wang, V. Leung, A. Stephenne, Interference-limited mixed Málaga-m and generalized-k dual-hop FSO/RF systems, in *2017 IEEE 28th Annual International Symposium on Personal, Indoor, and Mobile Radio Communications (PIMRC)* (2017), pp. 1–6
19. E. Balti, M. Guizani, Mixed RF/FSO cooperative relaying systems with co-channel interference. IEEE Trans. Commun. **66**(9), 4014–4027 (2018)
20. I.S. Ansari, F. Yilmaz, M.S. Alouini, Performance analysis of free-space optical links over $\mathcal{M}$´*a*laga ($\mathcal{M}$) turbulence channels with pointing errors. IEEE Trans. Wirel. Commun. **15**(1), 91–102 (2016)
21. H. AlQuwaiee, I.S. Ansari, M.S. Alouini, On the performance of free-space optical communication systems over double generalized gamma channel. IEEE J. Selec. Areas Commun. **33**(9), 1829–1840 (2015)
22. S.B. Bambiwal, A. Upadhya, R.S. Yaduvanshi, V.K. Pandey, Partial relay selection for combating the effects of co-channel interference in RF/FSO cooperative relaying. Opt. Commun. **475**, 126186 (2020)
23. A. Upadhya, V.K. Dwivedi, G.K. Karagiannidis, On the effect of interference and misalignment error in mixed RF/FSO systems over generalized fading channels. IEEE Trans. Commun. **68**(6), 3681–3695 (2020)
24. A. Upadhya, J. Gupta, V.K. Dwivedi, M.-S. Alouini, Impact of RF I/Q imbalance on interference-limited mixed RF/FSO TWR systems with non-zero boresight error. IEEE Wireless Commun. Lett. **10**(2), 416–420 (2021)
25. A. Upadhya, Investigation of mixed RF/FSO decode-and-forward NOMA cooperative relaying networks. Wireless Pers Commun **124**, 2923–2938 (2022)
26. A. Goel, R. Bhatia, Joint impact of interference and hardware impairments on the performance of mixed RF/FSO cooperative relay networks. Opt. Quant Electron, Springer **53**, 530 (2021)
27. A. Goel, R. Bhatia, Double relay selection for combating the impact of interference and hardware impairments in Mixed RF/FSO two way relay networks. Opt. Quant. Electron. **54**(11), 751 (2022)
28. M.O. Hasna, M.S. Alouini, A. Bastami, E.S. Ebbini, Performance analysis of cellular mobile systems with successive co-channel interference cancellation. IEEE Trans. Wirel. Commun. **2**(1), 29–40 (2004)
29. M.K. Simon, M.S. Alouini, *Digital Communication over Fading Channels*, 2nd edn. (Wiley, New York, 2005)
30. I.S. Gradshteyn, I.M. Ryzhik, *Table of Integrals, Series and Products* (Academic, New York, 2000)
31. I. Wolfram, Wolfram, Research, Mathematica Edition: Version 10.0. Wolfram Research, Inc., Champaign (2010)
32. P. Mittal, K. Gupta, An integral involving generalized function of two variables. Proc. Indian Acad. Sci. Section A **75**(3), 117–123 (1972)
33. H.R. Alhennawi, M.M.H. El Ayadi, M.H. Ismail, H.A.M. Mourad, Closed-form exact and asymptotic expressions for the symbol error rate and capacity of the H-Function fading channel. IEEE Trans. Veh. Technol. **65**(4), 1957–1974 (2016)
34. A. Kilbas, H-Transforms: Theory and Applications. Analytical Methods and Special Functions (Taylor and Francis, New York, 2004)
35. Z. Wang, G.B. Giannakis, A simple and general parameterization quantifying performance in fading channels. IEEE Trans. Commun. **51**, 1389–1398 (2003)
36. I.S. Ansari, S. Al-Ahmadi, F. Yilmaz, M.-S. Alouini, H. Yanikomeroglu, A new formula for the BER of binary modulations with dual-branch selection over Generalized-K composite fading channels. IEEE Trans. Commun. **59**, 2654–2658 (2011)
37. A. Annamalai, R.C. Palat, J. Matyjas, Estimating ergodic capacity of cooperative analog relaying under different adaptive source transmission techniques, in *Proceedings of IEEE Sarnoff Symposium* (2010), pp. 1–5
38. A. Lapidoth, S.M. Moser, M.A. Wigger, On the capacity of free space optical intensity channels. IEEE Trans. Inf. Theory **55**, 4449–4461 (2009)

Chapter 5
UAV-FSO Cooperative Relaying System with Generalized Interference Model

5.1 Introduction

The utilization of unmanned aerial vehicles (UAVs) has experienced a surge in significance, owing to their capability to maintain aerial stations and maneuver as needed, facilitating reliable communication with ground-based devices. This unique attribute has spurred extensive research interest, as UAVs emerge as viable candidates for establishing ad hoc networks. The UAVs can be stationed in the air and can be locomoted as per requirement for maintaining reliable communication with devices on the ground. This feature of the UAV has promoted a lot of research as due to this feature, UAVs are suited contenders for establishing reliable links. In addition to this, the flexibility of the UAV systems makes it an even more usable form of wireless communication, especially in topographies where it is not possible to obtain a LoS connection, a UAV can be set up in the air to establish a direct link with the communicating devices. It is only therefore that the future wireless generation standardization agencies are envisaging the UAV communication to connect millions of devices. Envisioned by future wireless generation standardization agencies, UAV communication holds promise in connecting millions of devices, leveraging multiple connections via millimeter-wave (mmWave) and FSO links to enhance bandwidth [1–7].

Assessing the performance of UAV communication systems necessitates accounting for various variables that dictate the overall QoS. Numerous research endeavors have delved into extensive analyses of UAV communication systems, employing modeling strategies such as 3D Cartesian coordinate axis representation, fading distribution, and path loss considerations, although often overlooking interference [8–14]. Nonetheless, the presence of interference can significantly impact multiple UAV systems, warranting its inclusion in performance evaluations [15–18].

Moreover, the secrecy aspects of UAV systems have been explored, highlighting concerns regarding unwanted signaling intrusions and emphasizing the importance

A. Upadhya et al., *Free-Space Optical Communication Systems for Next Generation Networks*, https://doi.org/10.1007/978-3-031-74805-9_5

of comprehensive analysis considering factors like LoS and NLoS modes of transmission [19]. By operating within the near-infrared or visible light spectrum, FSO systems capitalize on the favorable propagation characteristics of optical signals in free space, ensuring efficient data delivery over long distances. Moreover, it is imperative to establish a clear LoS between the transmitter and the receiver in FSO systems, as obstructions or adverse weather conditions can impede signal propagation and hinder communication reliability. Despite these challenges, the unparalleled bandwidth capabilities and potential for rapid deployment make FSO communication systems an attractive solution for various applications, including UAV-based communication networks and terrestrial point-to-point links. Integrating high-altitude platforms (HAPs) such as satellites with FSO systems further extends the realm of research, with analytical investigations conducted on performance metrics, considering turbulence models and fading conditions [20–22]. Similarly, physical layer security in satellite-FSO systems has been addressed, with closed-form expressions developed for secrecy outage probability under various relaying schemes and channel conditions [23]. Additionally, the impact of interference on satellite cognitive radio and proposals for hybrid RF/FSO relaying has been explored, incorporating threshold selection protocols to optimize link selection based on signal strength in [24, 25]. These diverse research endeavors collectively contribute to advancing the understanding and optimization of UAV and FSO-based communication systems in various operational scenarios.

Several findings have been reported in the past where the UAV system has been combined with FSO communication. Research work [26] investigates the optimization of UAV trajectories to maximize flight time and enhance the performance of FSO communication links. By integrating trajectory planning algorithms with FSO communication systems mounted on UAVs, the study aims to address the challenge of limited endurance and maximize the operational duration of UAV-based communication networks. The findings highlight the importance of considering both communication requirements and energy constraints in trajectory optimization, emphasizing the need for adaptive algorithms that dynamically adjust flight paths based on real-time environmental conditions and communication demands. Through simulation and experimental validation, the research demonstrates the effectiveness of trajectory optimization techniques in prolonging UAV flight time while maintaining reliable FSO communication links, thereby enhancing the overall efficiency and effectiveness of aerial communication systems. The paper [27] explores the integration of UAV with FSO communication networks employing an amplify-and-forward relaying mechanism. The study investigates the feasibility and performance of using UAVs as relay nodes to enhance signal strength and extend communication range in FSO links. By leveraging UAVs as aerial relays, the research aims to overcome LoS limitations and mitigate atmospheric turbulence effects, thereby improving the reliability and coverage of FSO communication systems. The findings demonstrate the effectiveness of the amplify-and-forward relaying scheme in compensating for signal attenuation and enhancing communication quality in UAV-assisted FSO networks. Through theoretical analysis and simulation studies, the paper provides insights into optimal relay placement, power allocation, and relay selection strategies to optimize system performance and achieve reliable data transmission over extended distances in challenging outdoor environments. The characteristics of the communication channel and optimizing parameters for FSO links with UAVs have been explained in detail in research work [28]. The study focuses on developing accurate channel models that capture the effects of atmospheric turbulence, motion dynamics, and other environmental factors on FSO communication performance. By considering parameters such as transmission power, beam divergence, and receiver aperture size, the research aims to optimize link quality and reliability in UAV-based FSO systems. Through theoretical analysis, simulation studies, and experimental validation, the paper provides insights into the impact of various channel parameters on system performance and proposes optimization strategies to enhance communication throughput and range. The findings contribute to the design and deployment of efficient UAV-based FSO communication networks for applications requiring high-speed data transmission over extended distances.

5.2 Wireless System Model

The UAVs can be used to connect to an FSO tower over a rooftop building. The LoS component of signal is very important in case of FSO systems. On the other hand, the UAVs have an inherent property of establishing a LoS link with the intended user. UAVs can therefore be used to connect to the FSO transmitter on a LoS link. Consider a case where there is no direct LoS between the ground user and the destination. In this case, the ground user can utilize a UAV for establishing a link using a dual-hop scheme. The UAV operates as an intermediate relay node between the ground station and the destination link. In a dense urban environment, when the FSO is used to provide large bandwidth backhaul, this model can be of immense importance. This prospect of the UAV is very helpful as it has the capability to maintain LoS for longer durations by changing the location. However, it is possible that there are several other UAVs that operate in the vicinity of the relay UAV. All interfering UAVs have impact on the performance of the system. The Cartesian coordinate system needs to be applied while modeling the UAVs as their locations along with height plays an important role in establishing the communication. The relay UAV will also act to convert the RF signal from the ground station into optical wireless signal which can be transmitted toward the destination FSO node on the rooftop of a high-rise building. The mode in which the relay UAV performs is also important. As per the cooperative relaying protocols, the preferred relaying strategies can be either DF or AF relays. If the relay node performs as a DF relay node, it receives signals from the ground station and then decrypts the message signal. Afterward, the relay node encodes the decoded version again and converts the signal in optical domain. The conversion can be done using LEDs or LASERs which can be driven by the RF signal from the ground station. The optical signal at the desired wavelength can be sent toward the destination node. The FSO aperture

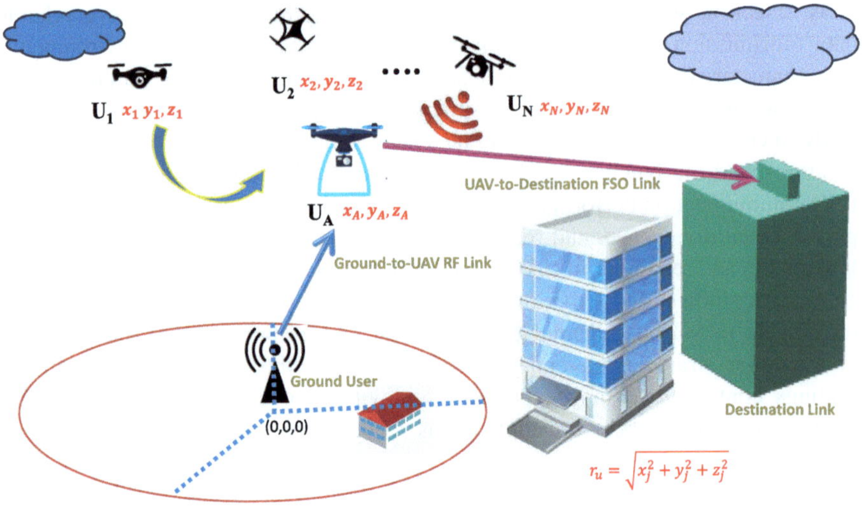

Fig. 5.1 UAV-FSO cooperative relaying strategy

will capture the signal and can send it toward the user at very high data rates. The relay UAV can enhance the reliability of the system by changing the location as per requirement for establishing a LoS path toward the FSO node.

The system model is shown in Fig. 5.1. The relay node has been assumed to operate as per the DF protocol. The ground user communicates with the FSO node through an intended UAV. This UAV receives the signal from the ground station, decodes it, and again encodes it before sending the same toward the FSO link. The intended UAV however functions in the presence of multiple other interfering UAVs. This reduces the reliability of the UAV as relay node although it can be used to maintain LoS with the FSO link. It has been assumed that accurate tracking system is used at the FSO node for eliminating any misalignment with the UAV. But a small pointing error has still been considered.

5.2.1 *UAV Channel Model*

We consider a similar system model as presented in [29] for the UAV channel. A ground user communicates with the UAV $\mathcal{U}_A$ on the uplink channel. For notation convenience, UAV $\mathcal{U}_A$ has been designated as the main UAV. This main UAV operates in the presence of N UAVs where the ith UAV is designated as $\mathcal{U}_i$. The UAV $\mathcal{U}_A$ has been placed at location denoted by (x_A, y_A, z_A), whereas the ith UAV has been assumed to be located in air at (x_i, y_i, z_i). One important consideration while determining the performance of the UAV is whether or not the LoS will be available. A formal notation for evaluating the probability of LoS has been presented in [14] which is as follows: $\xi_L(r_j) = \left[1 + \lambda \exp\left(-\delta\left(\frac{180}{\pi}\arctan(h_j/r_j) - \Lambda\right)\right)\right]^{-1}$ where

UAV height is represented by h_j and r_j with $j \in \{A, i\}$ denotes the distance of UAV, while the topology of the terrain depends on the values of λ and δ. The signal strength resultant of the power throughput from the UAV under the effect of the path loss can be written as [29, 30]

$$P_{b_j} = \begin{cases} \xi_N(r_j)\varpi_{N,j}h_{N,j}F_{N,j}{}^{-\alpha_N}, & \text{NLoS} \\ \xi_L(r_j)\varpi_{L,j}h_{L,j}F_{L,j}{}^{-\alpha_L}, & \text{LoS} \end{cases} \tag{5.1}$$

where $j \in \{A, i\}$ depending upon the UAV is main or interfering, and we define a constant $\varpi_{i,j} = N_j\eta_i$, where $i \in (N, L)$. This parameter depends on the radio frequency power that is transmitted by the UAV which has been included as N_j. The constant η_i represents the various extrinsic losses. The equivalent signal-to-interference-plus-noise ratio (SINR) can be written as

$$\Gamma_1 = \frac{\Lambda_U|h_u|^2}{\sum_{i=1}^{N}\Lambda_i|h_i|^2 + 1} \tag{5.2}$$

where

$$\Lambda_{yj} = \frac{P_y(r_y)\varpi_y F_y{}^{-\alpha_y}}{\sigma_n^2} \tag{5.3}$$

with $y \in (A, i)$, and additive white Gaussian noise (AWGN) power has been assumed to be σ_n^2. For deriving the outage probability, we derive a new closed-form expression for the PDF of the sum of the scaled Nakagami-m variable.

5.3 Sum of N Nakagami-m Random Variables

Consider a scaled Gamma random variable such that $Y = \Lambda X$. From the relationship of scaling of random variables, we have

$$f_Y(x) = \frac{1}{\Lambda} f_X\left(\frac{x}{\Lambda}\right) \tag{5.4}$$

The PDF for the ith scaled square Gamma random variable Y_i is given as

$$f_{Y_i}(x) = \frac{m_i^{m_i}}{\Lambda_i^{m_i}\Gamma(m_i)} x^{m_i-1}\exp\left(-\frac{m_i x}{\Lambda_i}\right) \tag{5.5}$$

Consider $Z = \sum_{i=1}^{N} Y_i$; then the moment generating function (MGF) of a random variable Y_1 is given as

$$E\left[e^{-sY_1}\right] = \int_0^\infty e^{-sx} f_{Y_i}(x)dx \tag{5.6}$$

The solution of the above integral gives

$$E\left[e^{-sY_1}\right] = \frac{m_1^{m_1}}{\Lambda_i^{m_1}\Gamma(m_1)}\left(s + \frac{m_1}{\Lambda_i}\right)^{-m_1} \tag{5.7}$$

Furthermore, the MGF of the sum of two square Gamma random variates can be obtained as

$$E\left[e^{-s(Y_1+Y_2)}\right] = E\left[e^{-sY_1}e^{-sY_2}\right] = E\left[e^{-sY_1}\right]E\left[e^{-sY_2}\right] \tag{5.8}$$

In obtaining the above relationship, the independence among the random variables Y_1 and Y_2 has been exploited. Using the description obtained in (5.7), the MGF $E\left[e^{-s(Y_1+Y_2)}\right]$ can be expressed as follows:

$$E\left[e^{-s(Y_1+Y_2)}\right] = \prod_{i=1}^{i=2} \frac{m_i^{m_i}}{\Lambda_i^{m_i}\Gamma(m_i)}\left(s + \frac{m_i}{\Lambda_i}\right)^{-m_i} \tag{5.9}$$

The method used for obtaining (5.9) can be extended in defining the MGF of sum of N Gamma random variates as follows:

$$E\left[e^{-s(Y_1+Y_2+\cdots+Y_N)}\right] = \prod_{i=1}^{i=N} \frac{m_i^{m_i}}{\Lambda_i^{m_i}\Gamma(m_i)}\left(s + \frac{m_i}{\Lambda_i}\right)^{-m_i} \tag{5.10}$$

The above expression for the MGF can be rewritten as

$$\mathcal{M}(s) = \sum_{q=1}^{N}\sum_{j=1}^{q} \frac{\Phi_q \mathcal{A}_{j,q}}{\Gamma(m_q)}\left(s + \Phi_q\right)^{-j} \tag{5.11}$$

where $\Phi_q = \frac{m_q^{m_q}}{\Lambda_q^{m_q}}$ and

$$\mathcal{A}_{j,q} = \begin{cases} \underset{s\to-\Theta_j}{\text{Lim}} \dfrac{\Phi_q\left(s+\Theta_j\right)}{\Gamma(m_q)} \prod_{j=1}^{N}\left(s+\Theta_j\right)^{-j}, & \text{Simple Pole} \\ \underset{s\to-\Theta_j}{\text{Lim}} \dfrac{\Phi_q\left(s+\Theta_j\right)^{j}}{\Gamma(m_q)} \prod_{j=1}^{N}\left(s+\Theta_j\right)^{-j}, & \text{Multiple Poles} \end{cases} \tag{5.12}$$

such that m_i can only belong to positive integers. The PDF of the sum of N scaled Gamma random variates can be established by taking the inverse Laplace transform of (5.10) and has been expressed as

$$f_{(Y_1+Y_2+\cdots+Y_N)}(x) = \sum_{q=1}^{N}\sum_{j=1}^{q} \frac{\Phi_q \mathcal{A}_{j,q}}{\Gamma(m_q)(j-1)!} x^{j-1} e^{-\Phi_q x} \tag{5.13}$$

It is emphasized that the above expression for the sum of scaled square Gamma random variables has not been included for analysis of UAV systems as per the limited knowledge of the authors. The derived expression can be utilized by researchers to accommodate nonidentical Nakagami-m random variables in the analysis. If we define the outage probability as the probability of $\Gamma_1 < x$, where x is the threshold SNR on the ground-to-UAV link, it is possible to obtain an expression for the outage probability using the following integral:

$$F_{\Gamma_1}(x) = \int_0^\infty F_{f_{|h_u|^2}}\left(\frac{x}{\Lambda_U}\Big(y+1\Big)\right) f_{(\sum_{i=1}^{N} A_i |h_i|^2)}(y)dy \tag{5.14}$$

The cumulative distribution function (CDF) for the UAV link has been assumed to follow the Nakagami-m fading model and is given as

$$F_{f_{|h_u|^2}}(x) = \frac{\gamma(m_u, x)}{\Gamma(m_u)} \tag{5.15}$$

The CDF and PDF can be taken from (5.15) and (5.13) to present the integral in (5.14) as [31]

$$F_{\Gamma_1}(x) = \sum_{q=1}^{N}\sum_{j=1}^{q} \frac{\Phi_q \mathcal{A}_{j,q}}{\Gamma(m_q)(j-1)!} \int_0^\infty \frac{\gamma(m_u, \left(\frac{x}{\Lambda_U}\Big(y+1\Big)\right))}{\Gamma(m_u)} y^{j-1} \exp\Big(-\Phi_q y\Big)dy \tag{5.16}$$

Utilizing identity for converting incomplete Gamma function into Meijer-G form, the above integral can be rewritten as follows:

$$F_{\Gamma_1}(x) = \sum_{q=1}^{N}\sum_{j=1}^{q} \frac{\Phi_q \mathcal{A}_{j,q}}{\Gamma(m_q)(j-1)!\Gamma(m_u)} \int_0^\infty y^{j-1} \mathrm{G}_{1,2}^{1,1}\left(\frac{x}{\Lambda_U}(y+1)\middle|\begin{matrix}1\\ m_u, 0\end{matrix}\right) \times \exp\Big(-\Phi_q y\Big)dy \tag{5.17}$$

where $\mathrm{G}_{p,q}^{m,n}\left(A\middle|\begin{matrix}a_p\\ b_q\end{matrix}\right)$ is the Meijer-G function defined in [32, Eq. (2.24.1)]. Substituting $\frac{x}{\Lambda_U}y = z$, the integral can be expressed as

$$F_{\Gamma_1}(x) = \sum_{q=1}^{N}\sum_{j=1}^{q} \frac{\Phi_q \mathcal{A}_{j,q}}{\Gamma(m_q)(j-1)!\Gamma(m_u)} \left(\frac{\Lambda_U}{x}\right)^{m_i} \int_0^{\infty} z^{j-1} \times \mathrm{G}_{1,2}^{1,1}\left(z + \frac{x}{\Lambda_U}\middle|\begin{matrix}1\\ m_u, 0\end{matrix}\right) \exp\left(-\frac{\Phi_q \Lambda_U z}{x}\right) dz \tag{5.18}$$

Invoking the identity [32, Eq. (2.24.3)], a closed-form solution to the integral in (5.18) can be presented as [31]

$$F_{\Gamma_1}(x) = \sum_{q=1}^{N}\sum_{j=1}^{q} \frac{\Phi_q \mathcal{A}_{j,q}}{\Gamma(m_q)(j-1)!\Gamma(m_u)} \sum_{k=0}^{\infty} \left(\frac{x}{\Lambda_U}\right)^{k-j} \frac{(-1)^k}{k!} \mathrm{G}_{3,3}^{2,2}\left(\frac{\Phi_q \Lambda_U}{x}\middle|\begin{matrix}k_1\\ k_2\end{matrix}\right) \tag{5.19}$$

where $k_1 = [1-j, 1+k-j-m_u, 1+k-j]$ and $k_2 = [0, k-j, 1+k-j]$. The outage probability expression is exact and can accept all valid possible combinations of the Nakagami-m parameter m_u. Although the result is in infinite series, the summation converges very fast and provides an exact value of outage probability. The outage probability expression takes into account for probability of NLoS, fading coefficient, 3D coordinates of the main, and the interfering UAV along with the number of interfering UAVs.

5.3.1 FSO Link

To bring variety into experience, another generalized optical scintillation statistical model is selected in this chapter. The Malaga fluctuation model has the ability to model optical signal properties from weak to moderate to strong turbulence conditions. The statistics of the optical wireless channel is modeled as a product of independent variables I_a and I_p as $I = I_a I_p$, where I_a denotes the atmospheric turbulence, whereas I_p accounts for the pointing errors. The atmospheric turbulence fading I_a on the FSO link is modeled as Malaga distribution. The PDF of γ_{FSO} on the FSO link can be expressed as [33]

$$f_{\Gamma_2}(\gamma) = \frac{\xi^2 \mathcal{M}_1}{2^r \gamma} \sum_{m=1}^{\beta} b_2 \mathrm{G}_{1,3}^{3,0}\left[\mathcal{M}_2 \gamma^{1/r}\middle|\begin{matrix}\xi^2+1\\ \xi^2, \alpha, m\end{matrix}\right] \tag{5.20}$$

where $\mathrm{G}_{p,q}^{m,n}\left[Arg\middle|\begin{matrix}a\\ b\end{matrix}\right]$ is the Meijer-G function defined in [34, Eq. (9.301)], r represents the type of optical demodulation scheme, constants M_1, M_2 are defined as $\mathcal{M}_1 = \frac{2\alpha^{\frac{\alpha}{2}}}{g^{1+\frac{\alpha}{2}}\Gamma(\alpha)}\left(\frac{g\beta}{g\beta+\Omega'}\right)^{\beta+\frac{\alpha}{2}}$, $\mathcal{M}_2 = \frac{\xi^2\alpha\beta(g+\Omega')}{[(\xi^2+1)(g\beta+\Omega')]\mu_r}$, and $b_1 =$

$\binom{\beta-1}{m-1}\frac{(g\beta+\Omega')^{1-\frac{m}{2}}}{\Gamma(m-1)}\left(\frac{\Omega'}{g}\right)^{m-1}\left(\frac{\alpha}{\beta}\right)^{\frac{m}{2}}$ [33]. The parameters involved in the Malaga distribution model are given as $g = 2b_0(1-\delta)$, $\Omega' = \Omega+2b_0\delta+2(2b_0\delta\Omega\cos(\phi_A-\phi_B))$. The pointing error parameters involved are $\xi = \frac{w_{zeq}}{2\sigma_s}$ [35], $w_{z_{eq}}^2 = \frac{w_z^2\sqrt{\pi}erf(v)}{2v\exp(-v^2)}$, $v = \frac{\sqrt{\pi}r}{2b_z}$ and $A_0 = [\text{erf(v)}]^2$, where w_z is the beam waist (calculated at e^{-2}) of the Gaussian spatial beam profile and $w_{z_{eq}}$ is equivalent beam waist at a distance of z.

$$F_{\Gamma_2}(\gamma) = \mathcal{M}_3 \sum_{m=1}^{\beta} b_3 \mathrm{G}_{r+1,3r+1}^{3r,1}\left[\mathcal{M}_4\gamma\middle|\begin{smallmatrix}1,\tau_1\\ \tau_2,0\end{smallmatrix}\right] \tag{5.21}$$

where $\mathcal{M}_3 = \frac{\xi^2\mathcal{M}_1}{2^r(2\pi)^{r-1}}$, $b_3 = b_2 r^{\alpha+m+1}$, $\mathcal{M}_4 = \frac{\mathcal{M}_2^r}{r^{2r}}$, with $\Delta(z : x)$ defined as $[\frac{x}{z}, \frac{x+1}{z}, \ldots, \frac{x+z-1}{z}]$.

5.4 Outage Probability Analysis

The outage probability of the proposed UAV-FSO scheme will depend on the performance of individual links. The overall SNR can be defined for DF protocol as

$$\Gamma_{eq} = \min(\Gamma_1, \Gamma_2) \tag{5.22}$$

As per the above definition, the outage probability can be defined as

$$\mathcal{P} = \Pr[\Gamma_{eq} < \gamma_{th}] \tag{5.23}$$

Therefore, the outage probability can be expressed as

$$\mathcal{P} = \Pr[\min(\Gamma_1, \Gamma_2) < \gamma_{th}] \tag{5.24}$$

According to the definitions of CDFs of the two links, the outage probability can be written as

$$\mathcal{P} = 1 - \Big((1 - F_{\Gamma_1}(\gamma_{th}))(1 - F_{\Gamma_2}(\gamma_{th}))\Big) \tag{5.25}$$

Substituting the requite distribution functions in the above, the closed-form expression for the outage probability can be derived as follows:

$$\mathcal{P} = 1 - \left(1 - \mathcal{M}_3 \sum_{m=1}^{\beta} b_3 \mathrm{G}_{r+1,3r+1}^{3r,1}\left[\mathcal{M}_4\gamma_{th}\middle|\begin{smallmatrix}1,\tau_1\\ \tau_2,0\end{smallmatrix}\right]\right)$$

$$\left(1 - \sum_{q=1}^{N}\sum_{j=1}^{q} \frac{\Phi_q \mathcal{A}_{j,q}}{\Gamma(m_q)(j-1)!\Gamma(m_u)}\right.$$
$$\left.\times \sum_{k=0}^{\infty}\left(\frac{\gamma_{th}}{\Lambda_U}\right)^{k-j}\frac{(-1)^k}{k!}\mathrm{G}_{3,3}^{2,2}\left(\frac{\Phi_q \Lambda_U}{\gamma_{th}}\middle|\begin{matrix}k_1\\k_2\end{matrix}\right)\right) \tag{5.26}$$

The exact outage probability expression represents a comprehensive model for UAV system, where all possible impacting factors have been taken into account. On the other hand, the FSO link factors like atmospheric turbulence, pointing errors, and type of optical demodulation scheme used also control the outage probability of the DF relay system.

5.5 Asymptotic Analysis

The high SNR approximate expressions can be developed from the closed-form expression provided in (5.26). The product of two CDFs involved in the definition of the outage probability can be neglected, and the approximate expression can be defined as

$$\mathcal{P} = \mathcal{M}_3 \sum_{m=1}^{\beta} b_3 \mathrm{G}_{r+1,3r+1}^{3r,1}\left[\mathcal{M}_4\gamma_{th}\middle|\begin{matrix}1,\tau_1\\ \tau_2,0\end{matrix}\right] + \sum_{q=1}^{N}\sum_{j=1}^{q}\frac{\Phi_q \mathcal{A}_{j,q}}{\Gamma(m_q)(j-1)!\Gamma(m_u)}$$
$$\times \sum_{k=0}^{\infty}\left(\frac{\gamma_{th}}{\Lambda_U}\right)^{k-j}\frac{(-1)^k}{k!}\mathrm{G}_{3,3}^{2,2}\left(\frac{\Phi_q \Lambda_U}{\gamma_{th}}\middle|\begin{matrix}k_1\\k_2\end{matrix}\right) \tag{5.27}$$

As the average electrical SNR on the FSO link increases, the value of $\mathcal{M}_4$ decreases, making the argument of the Meijer-G function become small. For this, an approximate result can be developed as follows:

$$\mathrm{G}_{r+1,3r+1}^{3r,1}\left[\mathcal{M}_4\gamma_{th}\middle|\begin{matrix}1,\tau_1\\ \tau_2,0\end{matrix}\right] \simeq \sum_{l=1}^{3r}\frac{\prod_{p=1}^{2}\Gamma(\tau_{2,p}-\tau_{2,l})\Gamma(\tau_{2,l})}{\prod_{p=2}^{r+1}\Gamma(\tau_{1,p}-\tau_{2,l})\Gamma(1+\tau_{2,l})}(\mathcal{M}_4\gamma_{th})^{\tau_{2,l}} \tag{5.28}$$

Similarly, as the average power from ground to UAV link increases, the argument of Meijer-G function increases, and it can be expressed as

$$\mathrm{G}_{3,3}^{2,2}\left(\frac{\Phi_q \Lambda_U}{\gamma_{th}}\middle|\begin{matrix}k_1\\k_2\end{matrix}\right)$$
$$\simeq \sum_{l=1}^{2}\frac{\prod_{p=1}^{2}\Gamma(k_{1,l}-k_{1,p})\prod_{p=1}^{2}\Gamma(1-k_{1,l}+k_{2,p})}{\Gamma(k_{1,l}-k_{2,3})\Gamma(1-k_{1,l}+k_{1,3})}\left(\frac{\Phi_q \Lambda_U}{\gamma_{th}}\right)^{k_{1,l}} \tag{5.29}$$

Using the above two representations of Meijer-G function, the asymptotic expression for the outage probability can be expressed as

$$\mathcal{P} = \mathcal{M}_3 \sum_{m=1}^{\beta} b_3 \sum_{l=1}^{3r} \frac{\prod_{p=1}^{2} \Gamma(\tau_{2,p} - \tau_{2,l})\Gamma(\tau_{2,l})}{\prod_{p=2}^{r+1} \Gamma(\tau_{1,p} - \tau_{2,l})\Gamma(1+\tau_{2,l})} (\mathcal{M}_4 \gamma_{th})^{\tau_{2,l}}$$
$$+ \sum_{q=1}^{N} \sum_{j=1}^{q} \frac{\Phi_q \mathcal{A}_{j,q}}{\Gamma(m_q)(j-1)!\Gamma(m_u)} \sum_{k=0}^{\infty} \left(\frac{\gamma_{th}}{\Lambda_U}\right)^{k-j} \frac{(-1)^k}{k!}$$
$$\times \sum_{l=1}^{2} \frac{\prod_{p=1}^{2} \Gamma(k_{1,l} - k_{1,p}) \prod_{p=1}^{2} \Gamma(1 - k_{1,l} + k_{2,p})}{\Gamma(k_{1,l} - k_{2,3})\Gamma(1 - k_{1,l} + k_{1,3})} \left(\frac{\Phi_q \Lambda_U}{\gamma_{th}}\right)^{k_{1,l}} \tag{5.30}$$

The above results suggest that the impact of all important variables has been included in the asymptotic expression. Due to this, the results are simpler, only in terms of elementary functions, and yet take into account for the fading on the UAV link, atmospheric turbulence, and pointing error on the FSO link.

5.6 Results

Some interesting results are demonstrated for the considered mixed UAV-FSO DF relaying systems. Figure 5.2 shows the impact of kind of optical demodulation scheme on FSO link and number of interfering UAVs near the intended UAV. One of the major contributions of the results in this chapter lies in the fact that the interfering signals have been modeled as unequal. Here, when $N = 2$ is considered, the INR that the two interfering UAVs provide has been selected as 1 dBm and 5 dBm. The Nakagami-m fading parameter for both the interfering UAVs has been assumed to be $m_i = 1$. The coordinates of the intended UAV has been assumed as $x_A, y_A, z_C = [20, 20, 20]$, while the two interfering UAVs are located at $x_1, y_1, z_1 = [30, 120, 30]$ and $x_2, y_2, z_2 = [30, 100, 70]$. The electrical SNR on the FSO link is considered as 30 dB. The outage probability has been depicted versus the average SNR on the ground to UAV link. It can be clearly seen that the outage probability decreases with increase in average SNR on the intended UAV link. However, the overall link is also dominated by the kind of optical demodulation scheme on the FSO link. The case when $N = 1$ and $r = 1$ provides the most favorable conditions as the coherent detection on the optical system with only one interfering UAV provides better link reliability.

On the other hand, Fig. 5.3 provides performance based on location of the intended UAV. For the purpose of demonstration, the location of UAV has been considered as $z_A = [20, 60]$, whereas $x_A, y_A = [20, 20]$. Therefore, the UAV is moved in vertical direction toward and away from the ground user. For each location of the UAV, the cases for $N = 1$ and $N = 2$ have been considered. Also, both

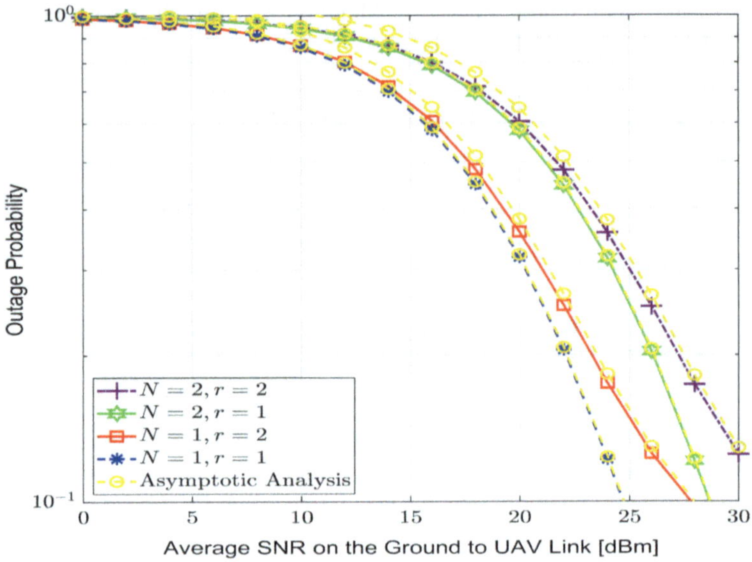

Fig. 5.2 Impact of kind of optical demodulation scheme on FSO link and number of interfering UAVs near the intended UAV

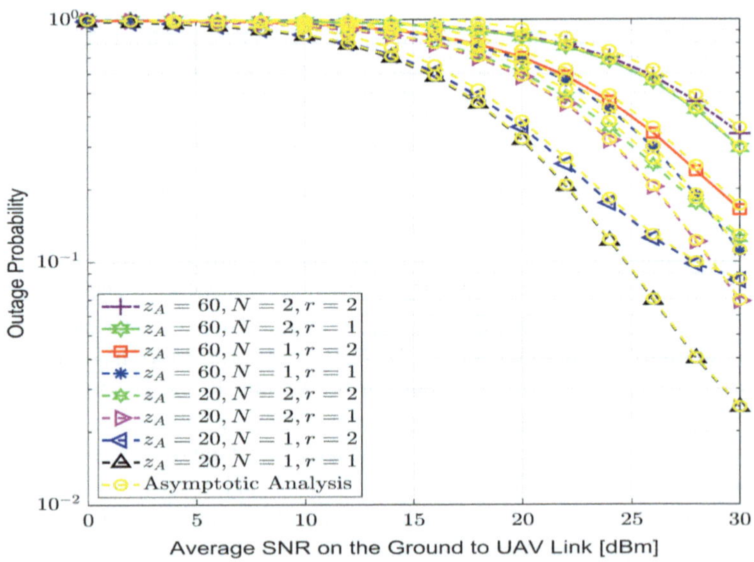

Fig. 5.3 Impact of location of the intended UAV along with interference and type of optical demodulation

coherent and IM/DD optical demodulation schemes have been considered in the analysis. It can be checked in the plot that when the UAV is near to the ground user, the outage probability is minimum. Nevertheless, even when UAV is nearer, the impact of interfering UAVs can be seen. Apart from that, $r = 1$, which corresponds to coherent demodulation, provides better outage performance. When $z_A = 20$, along with $N = 1, r = 1$ gives the minimum outage probability. For the same set of parameters, when IM/DD demodulation scheme is employed, the outage performance becomes worse. However, even if coherent demodulation is provided on the FSO link, when $N = 2$ interfering UAVs make the outage probability more than when IM/DD is employed. This means that the presence of interfering UAVs dominates the performance more compared to optical demodulation. Of course, the case when $N = 2, r = 2$ with $z_A = 20$ gives the maximum outage probability. However, all these outage probability values are still less compared to when $z_A = 60$. A similar behavior of outage probability in terms of number of interferers and type of optical demodulation can be seen in the plot. It is easy to conclude that the location of UAV can be used to mitigate interference from neighboring systems.

5.7 Conclusion

This chapter involved a relaying strategy where high-speed FSO link is connected to a ground user through UAV communication system. The DF relaying protocol has been considered, and the modeling of the UAV network considers small- and large-scale fading, the probability of LoS, 3D Cartesian of all the involved UAVs. Multiple unequal interfering UAVs have been taken into consideration. The atmospheric turbulence on the FSO link has been modeled using the Malaga distributed fading channel. With pointing error, the FSO link also takes into account for the kind of optical demodulation scheme that is used. For the considered system, a closed-form expression for the outage probability has been derived in exact form. This result has also been converted into asymptotic expression for high SNR. The results demonstrate that the interference impacts the outage probability of the link. However, the system model can be used as mode of interference mitigation by changing the location of the UAV. This system model will be beneficial for integrating UAV with high-speed large bandwidth FSO systems.

References

1. Y. Zeng, J. Lyu, R. Zhang, Cellular-connected UAV: Potential, challenges, and promising technologies. IEEE Wirel. Commun. **26**(1), 120–127 (2019)
2. N.H. Motlagh, T. Taleb, O. Arouk, Low-altitude unmanned aerial vehicles-based internet of things services: comprehensive survey and future perspectives. IEEE Internet Things J. **3**(6), 899–922 (2016)

3. W. Feng, J. Wang, Y. Chen, X. Wang, N. Ge, J. Lu, UAV-aided MIMO communications for 5G Internet of Things. IEEE Internet Things J. **6**(2), 1731–1740 (2019)
4. M. Mozaffari, W. Saad, M. Bennis, M. Debbah, Unmanned aerial vehicle with underlaid device-to-device communications: Performance and tradeoffs. IEEE Trans. Wirel. Commun. **15**(6), 3949–3963 (2016)
5. M.M. Azari, F. Rosas, K.-C. Chen, S. Pollin, Ultra reliable UAV communication using altitude and cooperation diversity. IEEE Trans. Commun. **66**(1), 330–344 (2018)
6. E. Turgut, M.C. Gursoy, Downlink analysis in unmanned aerial vehicle (UAV) assisted cellular networks with clustered users. IEEE Access **6**, 36313–36324 (2018)
7. Z. Hu, Z. Zheng, T. Wang, L. Song, Spectrum trading contract design for UAV assisted offloading in cellular networks, in *2018 IEEE International Conference on Communications (ICC)* (2018), pp. 1–6
8. M. Cui, G. Zhang, Q. Wu, D.W.K. Ng, Robust trajectory and transmit power design for secure UAV communications. IEEE Trans. Veh. Technol. **67**, 9042–9046 (2018)
9. Q. Wang, Z. Chen, H. Li, S. Li, Joint power and trajectory design for physical-layer secrecy in the UAV-aided mobile relaying system. IEEE Access **6**, 62849–62855 (2018)
10. C. Liu, J. Lee, T.Q.S. Quek, Safeguarding UAV communications against full-duplex active eavesdropper. IEEE Trans. Wirel. Commun. **18**(6), 2919–2931 (2019)
11. A. Li, Q. Wu, R. Zhang, UAV-enabled cooperative jamming for improving secrecy of ground wiretap channel. IEEE Wireless Commun. Lett. **8**(1), 181–184 (2018)
12. H. Lee, S. Eom, J. Park, I. Lee, UAV-aided secure communications with cooperative jamming. IEEE Trans. Veh. Technol. **67**(10), 9385–9392 (2018)
13. R.I. Bor-Yaliniz, A. El-Keyi, H. Yanikomeroglu, Efficient 3-D placement of an aerial base station in next generation cellular networks, in *2016 IEEE International Conference on Communications (ICC)* (2016), pp. 1–5
14. A. Al-Hourani, S. Kandeepan, S. Lardner, Optimal lap altitude for maximum coverage. IEEE Wireless Commun. Lett. **3**(6), 569–572 (2014)
15. W. Mei, R. Zhang, Aerial-ground interference mitigation for cellular-connected UAV. IEEE Wirel. Commun. **28**(1), 167–173 (2021)
16. A. Rahmati, S. Hosseinalipour, Y. Yapici, X. He, I. Guvenc, H. Dai, A. Bhuyan, Interference avoidance in uav-assisted networks: Joint 3d trajectory design and power allocation, in *2019 IEEE Global Communications Conference (GLOBECOM)* (2019), pp. 1–6
17. A. Fotouhi, H. Qiang, M. Ding, M. Hassan, L.G. Giordano, A. Garcia-Rodriguez, J. Yuan, Survey on UAV cellular communications: Practical aspects, standardization advancements, regulation, and security challenges. IEEE Commun. Surv. Tutorials **21**(4), 3417–3442 (2019)
18. H. Wang, J. Wang, J. Chen, Y. Gong, G. Ding, Network-connected UAV communications: Potentials and challenges. China Commun. **15**(12), 111–121 (2018)
19. X. Wang, H. Zhang, Y. Tian, V.C.M. Leung, Modeling and analysis of aerial base station-assisted cellular networks in finite areas under LOS and NLOS propagation. IEEE Trans. Wirel. Commun. **17**(10), 6985–7000 (2018)
20. E. Zedini, A. Kammoun, M.-S. Alouini, Performance of multibeam very high throughput satellite systems based on FSO feeder links with HPA nonlinearity. IEEE Trans. Wirel. Commun. **19**(9), 5908–5923 (2020)
21. R. Swaminathan, S. Sharma, A.S. MadhuKumar, *Performance Analysis of HAPS-Based Relaying for Hybrid FSO/RF Downlink Satellite Communication* (2020), pp. 1–5
22. R. Swaminathan, S. Sharma, N. Vishwakarma, A.S. Madhukumar, HAPS-based relaying for integrated space–air–ground networks with hybrid FSO/RF communication: A performance analysis. IEEE Trans. Aerosp. Electron. Syst. **57**(3), 1581–1599 (2021)
23. Y. Ai, A. Mathur, M. Cheffena, M.R. Bhatnagar, H. Lei, Physical layer security of hybrid Satellite-FSO cooperative systems. IEEE Photonics J. **11**, 1–14 (2019)
24. K. An, M. Lin, W.-P. Zhu, Y. Huang, G. Zheng, Outage performance of cognitive hybrid satellite–terrestrial networks with interference constraint. IEEE Trans. Veh. Technol. **65**(11), 9397–9404 (2016)

25. S. Shah, M. Siddharth, N. Vishwakarma, R. Swaminathan, A.S. Madhukumar, Adaptive-combining-based hybrid FSO/RF satellite communication with and without haps. IEEE Access **9**, 81492–81511 (2021)
26. J.-H. Lee, K.-H. Park, Y.-C. Ko, M.-S. Alouini, A UAV-mounted free space optical communication: trajectory optimization for flight time. IEEE Trans. Wirel. Commun. **19**(3), 1610–1621 (2020)
27. M.T. Dabiri, S. Khankalantary, M.J. Piran, I.S. Ansari, M. Uysal, W. Saad, C.S. Hong, UAV-assisted free space optical communication system with amplify-and-forward relaying. IEEE Trans. Veh. Technol. **70**(9), 8926–8936 (2021)
28. M.T. Dabiri, S.M.S. Sadough, M.A. Khalighi, Channel modeling and parameter optimization for hovering UAV-based free-space optical links. IEEE J. Sel. Areas Commun. **36**(9), 2104–2113 (2018)
29. A. Upadhya, On the reliability of interference limited unmanned aerial vehicles. Wireless Pers Commun. **129**, 119–131 (2023)
30. M. Alzenad, H. Yanikomeroglu, Coverage and rate analysis for unmanned aerial vehicle base stations with LOS/NLOS propagation, in *2018 IEEE Globecom Workshops (GC Wkshps)* (2018), pp. 1–7
31. A. Upadhya, A. Goel, Performance analysis of UAV communication system in the vicinity of multiple interferers, in *2023 9th International Conference on Signal Processing and Communication (ICSC)* (2023), pp. 135–138
32. O.I. Marichev A.P. Prudnikov, A. Brychkov, *Integrals and Series Special Functions* (CRC Press, New York, 1986)
33. I.S. Ansari, F. Yilmaz, M.S. Alouini, Performance Analysis of free-space optical links over $\mathcal{M}'a$laga ($\mathcal{M}$) turbulence channels with pointing errors. IEEE Trans. Wirel. Commun. **15**(1), 91–102 (2016)
34. I.S. Gradshteyn, I.M. Ryzhik, *Table of Integrals, Series and Products* (Academic, New York, 2000)
35. A.A. Farid, S. Hranilovic, Outage capacity optimization for free-space optical links with pointing errors. J. Lightwave Technol. **25**, 1702–1710 (2007).

Chapter 6
Impact of Interference on Mixed RF/MIMO-FSO Relaying Systems

6.1 Introduction

The present chapter analyzes the performance of interference limited mixed RF/MIMO-FSO communication system. With the aid of multiple apertures and receivers on the optical link, there is a possibility to enhance the performance of communication system. Although a cooperative relaying communication system can be used to enhance the quality, but the interference on the RF link also needs to be considered. Such a system model is very important given the high-speed requirement of future wireless communication systems. This chapter considers a system model where RF link is extended using the high-speed optical wireless link. To support more diversity on the FSO link, diversity is considered, which is modeled using the Gamma-Gamma turbulence statistical model. The RF link statistics for intended and the interfering signals has been modeled with the help of Nakagami-m distribution. An exact analysis for the RF link has been presented where the fading parameter m can take any value. Such a mathematical model will help to study the impact of various parameters with accuracy.

6.2 MIMO FSO Communication

FSO communication systems are being regarded as a promising solution for facilitating high-traffic communication in wireless systems, eliminating the necessity for physical fiber-optic cables. However, the optical wireless links are prone to atmospheric turbulence and pointing errors, scattering, and absorption, leading to potential degradation in signal quality and system performance limitations. There is a need to enhance the diversity order of the links. Such systems can be used to enhance the reliability of mixed RF/FSO cooperative relaying systems. To address these challenges, MIMO techniques are being utilized to leverage spatial diversity,

A. Upadhya et al., *Free-Space Optical Communication Systems for Next Generation Networks*, https://doi.org/10.1007/978-3-031-74805-9_6

thereby improving the reliability and overall performance of FSO connections. Various previous studies have analyzed the performance of multiple apertures in optical systems [1–8].

The implementation of MIMO techniques in FSO systems has shown notable enhancements in BER performance due to the exploitation of spatial diversity techniques. Spatial diversity necessitates the deployment of multiple transmit and receive optical apertures to establish a number of independent paths for the information signal, consequently mitigating the impacts of fading and turbulence. A study in [1] examined the BER performance of FSO transmission utilizing spatial diversity, divulging that spatial diversity-FSO systems effectively counter the adverse effects of optical signal fluctuations. The research illustrates that spatial diversity can reduce the BER by confirming that few transmission paths remain unaffected by deep fades, thereby preserving the integrity of the received signal. FSO channels commonly encounter Gamma-Gamma fading, which portrays the combined effects of small-scale and large-scale turbulence. MIMO techniques are especially adept at mitigating the influence of Gamma-Gamma fading. Research investigating the performance of a multi-aperture FSO system under Gamma-Gamma fading in [2] scenarios highlights the substantial improvement in system performance by offering multiple independent channels, which help in leveling out the impacts of turbulence. This advancement proves critical in upholding reliable communication links amid challenging atmospheric conditions.

The improved power efficiency of PPM has made it a modulation scheme commonly used in FSO systems. However, PPM systems are susceptible to fading and atmospheric disturbances, which can lead to high outages. The study in [3] explores the outage probability of Gaussian MIMO FSO channels with PPM scheme. The parallel optical beams that make multiple paths increase the likelihood that at least one channel will maintain a satisfactory SNR, as presented by the authors of the research work. This redundancy provided by MIMO significantly enhances the overall link reliability. Strong atmospheric turbulence can cause severe signal degradation in FSO systems. MIMO techniques can mitigate these effects through spatial diversity. In [4], the quality of optical wireless links with spatial diversity over strong atmospheric turbulence channels is examined. The research shows that MIMO FSO systems can achieve lower symbol error rates and higher link availability compared to single beam driven systems under strong turbulence conditions. This improvement is attributed to the diversity gain, which provides multiple independent paths for the signal, thereby reducing the probability of simultaneous deep fades across all channels.

Misalignment errors are modes of fading, which provide considerable obstacle for FSO systems. MIMO setups can mitigate the effects of these pointing errors by employing multiple transmit and receive openings. A study in [5] examines the performance of equal-gain combining (EGC) reception in FSO systems operating under composite-Gamma fading channels and pointing errors. Such an analysis is very cumbersome given the nature of the mathematically involved in the analysis. It is stumbled on that MIMO systems incorporating EGC can uphold an adaptable performance despite the presence of pointing errors. This resilience stems from the diversity gain that enables the system to counteract the effects of misalignment. Log-normal fading, which models the weak turbulence due to atmospheric conditions, is another critical factor affecting FSO systems. The log-normal fading model has been studied in [6] in the context of the MIMO technique with pointing errors. The results suggest that MIMO systems can enhance link robustness by averaging out the variations in signal strength across multiple channels, thereby improving the overall system performance. Incorporating advanced coding techniques with MIMO can further enhance the performance of FSO systems. In [7], a polar-coded MIMO FSO communication system over a Gamma-Gamma turbulence channel with spatially correlated fading is examined. The use of polar codes in conjunction with MIMO provides additional error correction capabilities, leading to significant improvements in BER performance and system reliability. Hybrid systems that combine RF and FSO technologies can benefit from MIMO techniques. In [8], the performance of a mixed RF/MIMO FSO variable gain dual hop transmission system is studied. The research shows that MIMO can improve the performance of hybrid systems by providing diversity gain in both the RF and FSO links, thereby enhancing the overall system robustness and reliability.

6.3 Motivation

In previous studies, the FSO channel is generally a SISO that does not provide any diversity order. Even if MIMO link is considered for an optical system, the constraint of complex mathematical analysis has restricted for extending the idea for cooperative relaying systems. However, given the fact that RF systems can be combined with FSO systems for enhanced service experience, it is important to consider the performance of the mixed RF/MIMO-FSO cooperative relays. With this, another important aspect is that of interference that needs attention. In fact, the modeling of the CCI with RF link can be appreciated in an exact manner that can reveal more about the performance of the systems. The study in the present chapter aims to analyze the mixed RF/MIMO-FSO cooperative relaying system in the presence of interference. The results that are derived in the chapter establish the fact the multiple apertures on the optical wireless links have the capability to enhance the user experience in wireless communication systems. More importantly, the results presented in the chapter also suggest that the interference plays a significant role while determining the performance of the mixed RF/MIMO-FSO cooperative relaying systems.

6.4 Mixed RF/MIMO FSO System Model

The chapter considers a system model in which communication between two nodes takes place through an intermediate relay node. The source node communicates with the relay node in the presence of RF interference near the relay device. The signal received on the relay node is expressed as

$$y_r = \sqrt{P_1} h_1 x_1 + \sum_{k=1}^{L} \sqrt{P_{I,k}} h_{I,k} x_{I,k} + n_1 \tag{6.1}$$

where the symbol x_1 is transmitted with power P_1 and the RF link fading statistics is represented as h_1. The AWGN that limits the RF link is denoted as n_1. The kth interferer transmits a power $P_{I,k}$ with fading coefficient $h_{I,k}$ to send interfering symbol $x_{I,k}$. As per the DF relay protocol, the relay node receives the signal and decodes the information symbol x_1 and retransmits the again encoded version $\hat{x}_1$ to the destination node.

The relay node has a function to convert the electrical signal from the first hop of communication into optical wavelength signals. In the present system model, that after this conversion is performed, the optical signals are passed through beam splitter. Such a device is an optical instrument that can convert one optical beam into two or more beams. The number of beams that can be generated depends on the angle of incidence of the beam that is to be split. Examples of such devices are prisms and mirrors. Mirror beam splitters are classified as either half-silvered or dielectric. These devices are used to convert a single optical beam into M beams. These then function as source of information from relay to destination link. At the destination end, the received optical signals, from M lasers, are imparted onto N photodetectors that convert the optical signal into an electrical signal. The received signal at the ith photodetector can be described as

$$y_d = \eta \sum_{m=1}^{M} I_{m,i} x_1 + n_{2,i}, \quad i = 1, 2, \ldots, N \tag{6.2}$$

where η denotes the optical-to-electrical conversion efficiency, I_m is the intensity from the mth Laser, while $n_{2,i}$ represents the AWGN component between the mth Laser and ith photodetector FSO link.

6.5 Outage Probability

The Nakagami-m distribution has been considered for modeling the small-scale fading on the RF link, the CDF of which can be given as

$$F_{\gamma_1}(x) = \frac{1}{\Gamma(m_1)} \gamma\left(m_1, \beta_1 x\right) \tag{6.3}$$

On the other hand, the interference on the RF link undergoes identically distributed Nakagami-m fading, the PDF for which is written as follows:

$$f_{\gamma_I}(x) = \frac{\beta_I^{m_I N}}{\Gamma(m_I N)} x^{m_I N - 1} \exp(-\beta_I x) \tag{6.4}$$

The effective SINR on the first hop of cooperative relaying can be defined as

$$\gamma_{RF} = \frac{\gamma_1}{\gamma_I + 1} \tag{6.5}$$

Based on the above definition, the overall CDF on the RF link can be defined as

$$F_{\gamma_{RF}}(x) = \int_0^\infty F_{\gamma_1}(x(y+1)) f_{\gamma_I}(y) dy \tag{6.6}$$

Plugging the expressions from the above definitions in (6.6), we have

$$F_{\gamma_{RF}}(x) = \frac{\beta_I^{m_I N}}{\Gamma(m_1)\Gamma(m_I N)} \int_0^\infty \gamma\Big(m_1, \beta_1(x(y+1))\Big) y^{m_I N - 1} \exp(-\beta_I y) dy \tag{6.7}$$

The incomplete Gamma function in the above expression can be represented in terms of Meijer-G function that redefines the integral in (6.7) as

$$F_{\gamma_{RF}}(x) = \frac{\beta_I^{m_I N}}{\Gamma(m_1)\Gamma(m_I N)} \int_0^\infty y^{m_I N - 1} \mathrm{G}_{1,2}^{1,1}\left[\beta_1(x(y+1)) \middle| \begin{matrix} 1 \\ m_1, 0 \end{matrix}\right] \exp(-\beta_I y) dy \tag{6.8}$$

In order to solve the above integral, the definition of the Meijer-G function in [9, Eq. (07.34.02.0001.01)] is expressed in the following:

$$\mathrm{G}_{1,2}^{1,1}\left[\beta_1(x(y+1)) \middle| \begin{matrix} 1 \\ m_1, 0 \end{matrix}\right] = \frac{1}{(2\pi i)} \int_{\mathcal{L}_\infty} \frac{\Gamma(m_1 - s_1)\Gamma(s_1)}{\Gamma(1+s_1)} (\beta_1(x(y+1)))^{s_1} ds_1 \tag{6.9}$$

Substituting from (6.9) into (6.8), we have

$$\begin{aligned} F_{\gamma_{RF}}(x) = {} & \frac{\beta_I^{m_I N}}{\Gamma(m_1)\Gamma(m_I N)} \frac{1}{(2\pi i)} \int_{\mathcal{L}_1} \frac{\Gamma(m_1 - s_1)\Gamma(s_1)}{\Gamma(1+s_1)} (\beta_1 x)^{s_1} \\ & \times \int_0^\infty y^{m_I N - 1} (y+1)^{s_1} \exp(-\beta_I y) dy ds_1 \end{aligned} \tag{6.10}$$

The inner integral is solved with the aid of identity provided in [10, Eq. (3.383.5)]. After this, confluent hypergeometric function can be expressed in terms of Meijer-G function using the relation [11, Eq. (4.46.1)]

$$F_{\gamma_{RF}}(x) = \frac{\beta_I^{m_I N}}{\Gamma(m_1)\Gamma(m_I N)} \frac{1}{(2\pi i)} \int_{\mathcal{L}_1} \frac{\Gamma(m_1 - s_1)\Gamma(s_1)}{\Gamma(1+s_1)} (\beta_1 x)^{s_1} \times \frac{1}{\Gamma(-s_1)} \mathrm{G}_{1,2}^{2,1}\left[\beta_I \middle| \begin{matrix} 1-m_I N \\ 0, -m_I N - s_1 \end{matrix}\right] ds_1 \tag{6.11}$$

Further, the contour integral definition of Meijer-G function as per [9, Eq. (07.34.02.0001.01)] is utilized to present the solution in terms of double contour integral as

$$F_{\gamma_{RF}}(x) = \frac{\beta_I^{m_I N}}{\Gamma(m_1)\Gamma(m_I N)} \frac{1}{(2\pi i)^2} \int_{\mathcal{L}_1} \int_{\mathcal{L}_2} \Gamma(-m_I N - s_1 - s_2) \times \frac{\Gamma(m_1 - s_1)\Gamma(s_1)}{\Gamma(1+s_1)\Gamma(-s_1)} \Gamma(-s_2)\Gamma(m_N + s_2)(\beta_1 x)^{s_1} {\beta_I}^{s_2} ds_1 ds_2 \tag{6.12}$$

To this end, the double contour integral is presented in the form of Fox-H function as given in [12, Eq. (1.1)]

$$F_{\gamma_{RF}}(x) = \frac{\beta_I^{m_I N}}{\Gamma(m_1)\Gamma(m_I N)} \mathrm{H}_{1,0:2,2:1,1}^{0,1:1,1:1,1} \times \left[\begin{matrix} (1+m_I N : -1, -1) \\ \text{---} \end{matrix} \middle| \begin{matrix} (1,1),(0,1) \\ (m_1,1),(0,1) \end{matrix} \middle| \begin{matrix} (1-m_I N, 1) \\ (0,1) \end{matrix} \middle| \beta_1 x, \beta_I \right] \tag{6.13}$$

The above-mentioned CDF expression for the RF link undergoing interference is new and considers the effect of interference and small-scale fading on the RF link.

On the other hand, the irradiance $I_{m,i}$ follows the Gamma-Gamma distribution as given below [13]:

$$f(I) = \frac{2(\alpha\beta)^{\frac{\alpha+\beta}{2}}}{\Gamma(\alpha)\Gamma(\beta)} I^{\frac{\alpha+\beta}{2}-1} K_{\alpha-\beta}\left(2(\alpha\beta I)^{\frac{1}{2}}\right) \tag{6.14}$$

where the α and β are related to scintillation index S.I. as S.I. $= \frac{1}{\alpha} + \frac{1}{\beta} + \frac{1}{\alpha\beta}$. Recently, the PDF of MIMO-FSO turbulence channel has been presented in [14] as

$$f_{\gamma_{FSO}}(x) = \frac{(\alpha_s\beta_s)^{(\alpha_s+\beta_s)/2} x^{(\alpha_s+\beta_s)/4-1}}{2\Gamma(\alpha_s)\Gamma(\beta_s)\bar{\gamma}_{FSO}^{(\alpha_s+\beta_s)/4}} \mathrm{G}_{0,2}^{2,0}\left[\alpha_s\beta_s\left(\frac{x}{\bar{\gamma}_{RD}}\right)^{\frac{1}{2}} \middle| \begin{matrix} \text{---} \\ \frac{\alpha_s-\beta_s}{2}, \frac{\beta_s-\alpha_s}{2} \end{matrix}\right] \tag{6.15}$$

where γ_{FSO} represents the instantaneous SNR, while $\bar{\gamma}_{FSO}$ denotes the average SNR on the wireless optical link with diversity. In the above equation, the instantaneous SNR can be presented as $\gamma_{FSO} = \bar{\gamma}_{FSO} I^2$, where $I = \sum_{n=1}^{N}\sum_{m=1}^{M} I_{mn}$, and I_{mn} denotes the atmospheric turbulence of the FSO link between mth transmitter and nth optical receiver. Moreover, G[.] denotes the Meijer's G function [10, Eq. (9.301)], $\Gamma(.)$ denotes the Gamma function, and $\alpha_s = MN\alpha + \epsilon$, $\beta_s = MN\beta$. The PDF parameters α, β and $\epsilon = (MN - 1)\frac{-0.127-0.95\alpha-0.0058\beta}{1+0.00124\alpha+0.98\beta}$ are used to describe the turbulence experienced by signals. The analytical expression for the CDF of diversity aided FSO link over Gamma-Gamma distribution is given as [14]

$$F_{\gamma_{FSO}}(x) = A_1 \mathrm{G}_{1,5}^{4,1}\left[\frac{A_2}{\bar{\gamma_{RD}}}x \middle| \begin{matrix} 1 \\ k_1, 0 \end{matrix}\right] \tag{6.16}$$

where $A_1 = \frac{2^{\alpha_s+\beta_s-2}}{\pi\Gamma(\alpha_s)\Gamma(\beta_s)}$, $A_2 = \frac{(\alpha_s\beta_s)^2}{16}$, $k_1 = \frac{\alpha_s}{2}, \frac{\alpha_s+1}{2}, \frac{\beta_s}{2}, \frac{\beta_s+1}{2}$.

As per the DF relaying strategy, the probability of outage can be defined as the event when the end-to-end SINR becomes smaller than the predefined threshold SNR. If the equivalent SINR is defined as $\gamma_{eq} = \min(\gamma_{RF}, \gamma_{FSO})$, then the outage probability can be expressed as $Pr[\gamma_{eq} < \gamma_{th}]$. With this definition, the outage probability of the considered system model can be derived using the results in (6.16) and (6.13)

$$\begin{aligned} P_{out} &= \left(1 - \frac{\beta_I^{m_I N}}{\Gamma(m_1)\Gamma(m_I N)} \mathrm{H}_{1,0:2,2:1,1}^{0,1:1,1:1,1}\right. \\ &\times \left.\left[\begin{matrix} (1+m_I N : -1, -1) \\ \text{---} \end{matrix}\middle| \begin{matrix} (1,1),(0,1) \\ (m_1,1),(0,1) \end{matrix}\middle| \begin{matrix} (1-m_I N, 1) \\ (0,1) \end{matrix}\middle| \beta_1\gamma_{th}, \beta_I\right]\right) \\ &\times \left(1 - A_1 \mathrm{G}_{1,5}^{4,1}\left[\frac{A_2}{\bar{\gamma_{RD}}}\gamma_{th}\middle| \begin{matrix} 1 \\ k_1, 0 \end{matrix}\right]\right) \end{aligned} \tag{6.17}$$

The derived expression is an exact analytical result for the considered system model. The expression is applicable for non-integer values of both Nakagami-m parameters, i.e., m_1 and m_I. The closed-form expression considers the effect of amount of fading on the RF link, turbulence on the optical link with pointing errors, and multiple apertures on the optical link along with interference on the RF link.

6.6 Results

The improvement from the increase in diversity order on the wireless optical link is demonstrated in Fig. 6.1. The threshold SNR for the study has been assumed to be $\gamma_{th} = -2$ dB. On the RF link, the fading has been modeled using $m_1 = 3/2$. It has been considered that there are $L = 2$ interferers on the RF link with interfering INR of 1 dB and fading coefficient of $m_I = 2$. The FSO link has been modeled using $\alpha = 2.902$, $\beta = 2.51$, which corresponds to weak turbulence. The diversity orders considered in the analysis are $M \times N = [1 \times 1, 2 \times 1, 2 \times 2]$. In general, the outage probability reduces with increase in average SNR per hop. Specifically, the outage probability reduces with increase in the order of diversity. The plot also

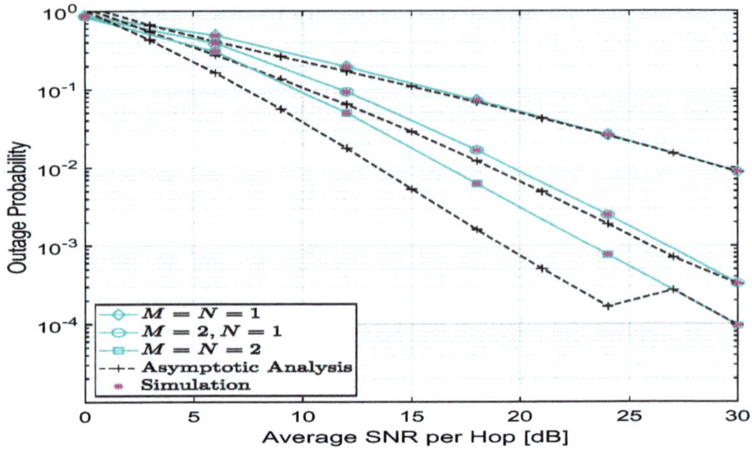

Fig. 6.1 Outage probability analysis for increase in diversity order on the optical wireless link

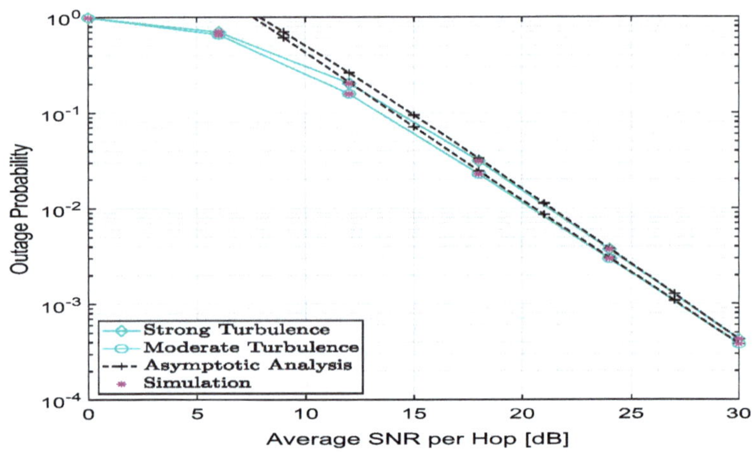

Fig. 6.2 Outage probability analysis for effect of turbulence on the optical wireless link

presents the asymptotic high SNR approximate expression of the system model. The results are also verified using the Monte Carlo simulations. The main inference that can be taken from the analysis is that the number of optical transmitter and receiver apertures can be increases to enhance the reliability of the network.

The analysis for change in strength of turbulence on optical communication link is depicted in Fig. 6.2. The threshold SNR in the study has been considered to be $\gamma_{th} = 2$ dB. It has been assumed that there are $L = 2$ interference sources on the RF link. The strength of turbulence has been modified on the optical link by choice of turbulence parameters α, β. The combination of model parameter for representing

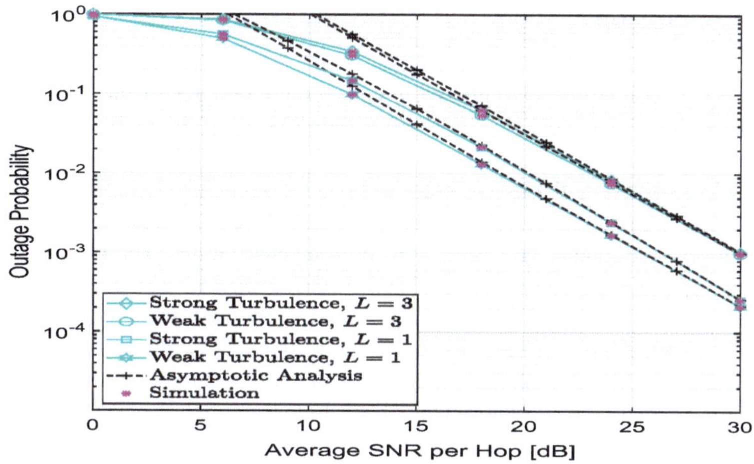

Fig. 6.3 Study of outage probability for varying strength of turbulence and the number of interferers on the RF link

strong turbulence is $\alpha, \beta = (2.064, 1.342)$. On the other hand, moderate turbulence considered using model parameter combination of $\alpha, \beta = (2.296, 2.51)$. The plot also demonstrates the high SNR asymptotic expression on the same axis. The exact results converge with the approximate results in the high SNR regime. As it is evident in the plot, the moderate turbulence conditions on the optical wireless link show lower outage probability. These results are important considering the fact that even in the case of designing an MIMO transceiver on the FSO link, the impact of adverse effect on magnitude and phase changes on the optical signal needs to be considered.

The study of effect of interference on the outage probability with respect to the strength of atmospheric turbulence has been depicted in Fig. 6.3. The number of interferers on the RF link has been considered as $L = [1, 3]$, while strong and weak turbulence has been considered on the optical wireless link. Each interfering device has been assumed to provide an INR of $\bar{\gamma}_I = 2$ dB. Also, the fading coefficient on the RF link has been considered as $m_1 = 2$, while the fading coefficient on the interfering links has been taken as $m_I = 3/2$. It can be seen in the plot that the number of interfering devices in the vicinity of the relay node makes a significant contribution. It can be checked in the results that when there are $L = 3$ such devices, the effect of strength of turbulence is comparatively less. This is because even if weak turbulence, the outage probability curve is nearer to the strong turbulence when devices are more. When the interfering devices become $L = 1$, the outage probability is very less for both strong and weak turbulence conditions. The plot also demonstrates the asymptotic and Monte Carlo simulation results.

6.7 Conclusions

The diversity techniques have proved to enhance the efficacy of the wireless communication networks in the past. For radio communication channels, the concept of MIMO has existed from past several decades. The same technique of establishing multiple links between the transmitters and receivers can also be utilized in optical wireless communication networks. The multiple links can be used to multiplex the data or to use redundant data streams to fight against fading. The present chapter presents highlights of the mixed RF/MIMO-FSO cooperative relaying systems in the presence of interference. The strategy used in the system is the way forward for contemplating the requirements of the future wireless communication systems. The signals on the RF link have been assumed to undergo fading that follows Nakagami-m distribution. The MIMO FSO link is modeled using the Gamma-Gamma turbulence model. The performance is significantly improved using spatial diversity on the FSO link. However, the interference on the RF link also serves poorly. The results in the chapter show that when the interfering powers near the relay node increase, there is degradation in the reliability of the link. This suggests that the link budget of the mixed RF/MIMO-FSO systems must also include the impact of interference.

References

1. S.M. Navidpour, M. Uysal, M. Kavehrad, BER performance of free-space optical transmission with spatial diversity. IEEE Trans. Wirel. Commun. **6**(8), 2813–2819 (2007)
2. E. Bayaki, R. Schober, R.K. Mallik, Performance analysis of MIMO free-space optical systems in gamma-Gamma fading. IEEE Trans. Commun. **57**(11), 3415–3424 (2009)
3. N. Letzepis, A.G. I Fabregas, Outage probability of the Gaussian MIMO free-space optical channel with PPM. IEEE Trans. Commun. **57**(12), 3682–3690 (2009)
4. T.A. Tsiftsis, H.G. Sandalidis, G.K. Karagiannidis, M. Uysal, Optical wireless links with spatial diversity over strong atmospheric turbulence channels. IEEE Trans. Wirel. Commun. **8**(2), 951–957 (2009)
5. N.I. Miridakis, T.A. Tsiftsis, EGC reception for FSO systems under mixture-gamma fading channels and pointing errors. IEEE Commun. Lett. **21**(6), 1441–1444 (2017)
6. M.T. Dabiri, M.J. Saber, S.M.S. Sadough, On the performance of multiplexing FSO MIMO links in log-normal fading with pointing errors. IEEE/OSA J. Opt. Commun. Networking **9**(11), 974–983 (2017)
7. L. Han a, H. Jiang, Y. You, Z. Ghassemlooy, On the performance of a mixed RF/MIMO FSO variable gain dual-hop transmission system. Elsevier/Optics Commun. **420**, 59–64 (2018)
8. J. Fang, M. Bi, S. Xiao, G. Yang, L. Liu, Y. Zhang, W. Hu, Polar-coded MIMO FSO communication system over Gamma-Gamma turbulence channel with spatially correlated fading. IEEE/OSA J. Opt. Commun. Networking **10**(11), 915–923 (2018)
9. I. Wolfram, Wolfram, research, mathematica edition: Version 10.0. Champaign (Wolfram Research, Inc., New York, 2010)
10. I.S. Gradshteyn, I.M. Ryzhik, *Table of Integrals, Series and Products* (Academic, New York, 2000)

11. O.I. Marichev, A.P. Prudnikov, A. Brychkov, *Integrals and Series Special Functions* (CRC Press, New York, 1986)
12. P. Mittal, K. Gupta, An integral involving generalized function of two variables. Proc. Indian Acad. Sci. Section A **75**(3), 117–123 (1972)
13. M. Al-Habash, L.C. Andrews, R.L. Phillips, Mathematical model for the irradiance probability density function of a laser beam propagating through turbulent media. Opt. Eng. **40**(8), 1554–1562 (2001)
14. D.A. Luong, A.T. Pham, Average capacity of MIMO free-space optical gamma-gamma fading channel. IEEE Int. Conf. Commun. ICC, 2014 **65**(4), 3354–3358 (2014)

Chapter 7
Noma Aided Interference Limited Mixed RF/FSO Relaying

7.1 Background

In this chapter, we present a non-orthogonal multiple access (NOMA) scheme to evaluate the performance of a mixed RF/FSO cooperative relaying system. The relay node within this hybrid RF/FSO framework operates under the DF protocol. Additionally, it is assumed that a direct communication link exists between the source and destination nodes, enhancing the system's reliability. The RF links are characterized using the Nakagami-m distribution, which accurately models the small-scale fading in RF communication channels. For the FSO links, the atmospheric turbulence effects are described using the Malaga distribution, a comprehensive model that captures a wide range of turbulence conditions. The FSO link model employed in this analysis also accounts for the impact of pointing errors and includes two types of optical demodulation schemes: IM/DD and coherent demodulation.

For the proposed NOMA network, we have derived analytical expressions for the outage probability and ergodic capacity. The outage probability analysis for the mixed RF/FSO DF system has been further extended to the high SNR regime, allowing the results to be expressed in elementary functions. This analytical framework facilitates a deeper understanding of how various RF and FSO system parameters influence the overall reliability and throughput of the system. Moreover, the numerical results have been rigorously validated through Monte Carlo simulations. These simulations confirm the accuracy of the analytical expressions and provide insights into the practical performance of the proposed mixed RF/FSO NOMA system under different operating conditions. This comprehensive analysis highlights the potential of NOMA in enhancing the efficiency and capacity of hybrid communication systems. By addressing the unique challenges posed by both RF and FSO links, this work contributes to the advancement of robust and high-performance wireless communication technologies.

A. Upadhya et al., *Free-Space Optical Communication Systems for Next Generation Networks*, https://doi.org/10.1007/978-3-031-74805-9_7

7.2 Introduction

The increasing demand for bandwidth in RF wireless communication systems has driven researchers to explore FSO communication systems as a viable alternative. Cooperative relaying systems (CRS) have garnered significant interest due to their potential to leverage the advantages of mixed RF/FSO relaying systems [1, 2]. However, conventional dual hop CRS are often criticized for spectrum inefficiency, as they require two time slots for transmission. Non-orthogonal multiple access (NOMA) schemes have gained attention for their ability to allow multiple users to share the same time–frequency resources, thereby enhancing spectral efficiency [3–12]. Integrating CRS with NOMA to form mixed RF/FSO systems can effectively combine the benefits of both RF and FSO technologies, addressing the spectrum inefficiency issue of traditional CRS. In the context of conventional RF NOMA systems, Ding et al. in [4] have utilized the spectral efficiency advantage of NOMA for short-range CRS. The primary motivation behind combining NOMA with CRS, as suggested in [4], is that the near user can decode the transmitted symbol using successive interference cancelation (SIC) and subsequently act as a relay node within the CRS framework.

For performance analysis of NOMA systems, Poisson clustering has been adopted in [5]. Meanwhile, the authors of [6] have investigated the performance of full-duplex device-to-device (D2D) communication systems operating on the NOMA principle. In [7], the switching between full-duplex and half-duplex modes for a DF NOMA system has been analyzed. This study also considers scenarios with and without a direct link between the base station and the far user. Additionally, research in [8] has focused on improving the performance of cell-edge users in a two-user NOMA system. On the other hand, an ON/OFF mechanism is studied to determine whether cooperative communication should be established. Analytical results for outage probability and sum rate have been used to compare the performance of the proposed schemes in [8].

The operation of NOMA for FSO systems has been explored extensively in recent literature [9–12]. In [11], an error rate analysis of an FSO-aided MISO NOMA system is conducted. This study establishes that the impact of SIC can be minimized by employing a 1-bit feedback-based beamforming scheme. In [9], NOMA is presented as a viable option for 5G backhauling, where the FSO link turbulence is modeled using the Gamma-Gamma fading model. The results for outage probability in this study are provided in the form of integrals evaluated numerically.

The performance of a mixed RF/FSO AF relaying network is investigated in [10], where the destination node utilizes dynamic order decoding to prioritize users. In a recent study [12], two RF users transmit information to the destination node through an AF relay system in the presence of multiuser interference. This work derives closed-form expressions for the outage probability, considering a backhaul FSO link subjected to Gamma-Gamma turbulence and pointing errors. Besides outage analysis, [12] also provides a tractable form for the ergodic capacity analysis and extends this analysis to conventional RF NOMA systems.

From the introduction of mixed RF/FSO cooperative relaying systems in [2], several research works including but not limited to [13–22] have analyzed the performance of these systems. Relay-aided AF and DF relaying systems over $\alpha - \mu$ turbulence channels have been presented in [13] and [14], respectively. The impact of interference on mixed RF/FSO two-way AF relaying scheme where the first hop is implanted with user diversity has been presented in [15]. Significantly, nonzero boresight component of pointing error, which is generally excluded in analysis, has been considered by the authors of [16], where the FSO link has been modeled using the $\alpha - \mu$ distribution, whereas the RF link has been modeled using the $\eta - \mu$ fading model. Moreover, two forms of AF relaying protocols, namely fixed and variable gain, have been addressed for mixed RF/FSO cooperative relaying systems where the FSO link has been modeled using the Malaga distribution by the authors of [17]. Further, the authors of [18] have analyzed the impact of interference on the reliability of mixed RF/FSO DF relaying system. Performance analysis of mixed RF/FSO AF relaying systems over double generalized Gamma channels has been performed by the authors of [19]. Closed-form expressions for the outage probability, bit error rate (BER), and ergodic capacity have been derived in [19]. Relay selection (PRS) based on outdated CSI has been advocated to combat the impact of co-channel interference in research work [20] where the FSO channel has been modeled with the help of double generalized Gamma turbulence channel with pointing error.

It is important to note that while the literature [3–8] focuses on NOMA within the RF domain, the research works [9–12] implement NOMA in FSO communication. Notably, [11] pertains to the analysis of MISO NOMA systems, whereas mixed RF/FSO CRS for NOMA have been examined in [9, 10, 12]. Importantly, the expressions for outage probability in [9] and for ergodic capacity in [12] are provided only in integral form, lacking closed-form solutions.

Additionally, the studies [9, 10, 12] utilize the Gamma-Gamma model to represent atmospheric scintillation on the FSO link and employ the AF relaying strategy at the relay node. However, the DF relaying strategy has been shown to outperform AF protocols [23]. Furthermore, the Malaga turbulence model, which encompasses the Gamma-Gamma model as a special case, is better suited for characterizing irradiance fluctuations under weak-to-moderate-to-strong turbulence conditions [24]. It is also significant that the RF links in the mixed RF/FSO relaying systems discussed in [9, 10, 12] are modeled using the Rayleigh fading model. The Nakagami-m model, which represents a broader range of fading scenarios including Rayleigh fading as a limiting case, could provide a more comprehensive representation of multipath components.

Motivated by these considerations, the present research investigates the performance of a mixed RF/FSO DF relaying system with NOMA implemented between the source, relay, and destination nodes. In this study, the RF link is modeled using the Nakagami-m distribution, and the irradiance fluctuations on the FSO link are characterized by the Malaga distribution. For this system, we derive closed-form expressions for both the outage probability and the ergodic capacity. Furthermore, the closed-form expression for the outage probability is extended to derive asymptotic approximations for high SNR scenarios. The results and corresponding analysis presented in the present chapter have been taken from the study published in [25]. The system model that has been considered in this chapter is very practical as the combination of FSO and NOMA can be a solution to small cell structure in the future wireless communication systems.

7.3 System Model

Consider a DF cooperative NOMA system, where a source node S communicates with the destination node D through a relay node R and also via a direct link between the source and destination. During the first time slot, the source transmits to the relay over an RF link. The signal received at the relay node R can be expressed as

$$y_{12} = h_{12}(\sqrt{a_1 P_1} s_1 + \sqrt{a_2 P_2} s_2) + n_{12} \tag{7.1}$$

where $n_{12} \backsim \mathcal{N}(0, \sigma_{n1}^2)$ represents the additive white Gaussian noise (AWGN) on the source-to-relay link with zero mean and variance σ_{n1}^2 and h_{12} denotes the fading coefficient of this link. The two symbols s_1 and s_2 are sent using power coefficients of a_1 and a_2 and using power P_1 and P_2, respectively. Importantly, as per the NOMA protocol, $a_1 + a_2 = 1$, and it is assumed that $a_1 > a_2$ [3]. Moreover, the relay node decodes s_1 considering s_2 as noise and acquires s_2 it using SIC to eliminate s_1. The symbol s_2 received at the destination node over the FSO link can be written as

$$y_{23} = \left(g_{22}\sqrt{P_F}\right)^{\frac{r}{2}} s_2 + +n_{23} \tag{7.2}$$

where g_{22} denotes the FSO link irradiance, P_F is optical power transmitted, and $n_{23} \backsim \mathcal{N}(0, \sigma_{n2}^2)$. The parameter r denotes the type of optical demodulation scheme where $r = 1$ corresponds to coherent, while $r = 2$ represents intensity modulation, direct detection. However, in the considered NOMA scheme, the source node S also sends both symbols to the destination node D over the direct RF link during the first time slot. The signal received at the destination node in this first time slot can be expressed as

$$y_{13} = h_{13}(\sqrt{a_1 P_1} s_1 + \sqrt{a_2 P_2} s_2) + n_{13} \tag{7.3}$$

where the fading coefficient on the RF $S - D$ link has been denoted by h_{13} with AWGN factor of $n_{13} \backsim \mathcal{N}(0, \sigma_{n3}^2)$.

To this end, on the $S - R$ link in the first time slot, the SNRs for s_1 and s_2 at the relay node are, respectively, given as

$$\gamma_{12,s1} = \frac{|h_{12}|^2 a_1 P_1}{|h_{12}|^2 a_2 P_2 + \sigma_{n1}^2} \tag{7.4}$$

and

$$\gamma_{12,s2} = \frac{|h_{12}|^2 a_2 P_2}{\sigma_{n1}^2}. \tag{7.5}$$

In the same time slot, the destination node decodes the symbol s_1 on the $S-D$ link, the SNR for which can be expressed as

$$\gamma_{13,s1} = \frac{|h_{13}|^2 a_1 P_1}{|h_{12}|^2 a_2 P_2 + \sigma_{n3}^2} \tag{7.6}$$

Moreover, from (7.2), the SNR at the destination node can be expressed as

$$\gamma_{23,s2} = \frac{|g_{23}|^2 P_F}{\sigma_{n2}^2} \tag{7.7}$$

Assuming that $\lambda_{12} = \frac{|h_{12}|^2 P_1}{\sigma_{n1}^2}$ and $\lambda_{13} = \frac{|h_{13}|^2 P_1}{\sigma_{n3}^2}$, the SNR statistics on the RF link has been modeled using the Nakagami-m fading model, the probability density function (PDF) of which is given as

$$f_{\lambda_{1j}}(x) = \frac{\beta_{1j}^{m_{1j}}}{\Gamma(m_{1j})} x^{m_{1j}-1} \exp\Big(- \beta_{1j} x\Big) \tag{7.8}$$

where m_{1j} denotes the Nakagami-m fading parameter and $\beta_{1j} = \frac{m1j}{\bar{\gamma}_{1j}}$, with $j = \{2, 3\}$ and $\bar{\gamma}_{1j} = \mathbb{E}\left[\frac{|h_{1j}|^2 P_j}{\sigma_{n1}^2}\right]$ denoting the average SNR on the RF links. On the other hand, the optical wireless channel on the $R-D$ link has been modeled using the Malaga distribution, the PDF and cumulative distribution function (CDF) of which is given respectively as [17]

$$f_{\gamma_{23,s2}}(x) = \frac{\xi^2 A}{x 2^r} \sum_{m=1}^{\delta} b_m \mathrm{G}_{1,3}^{3,0}\left(B\Big(\frac{x}{\mu_r}\Big)^{1/r} \middle| \begin{matrix} \xi^2+1 \\ \xi^2, \alpha, m \end{matrix} \right) \tag{7.9}$$

and

$$F_{\gamma_{23,s2}}(x) = C \sum_{m=1}^{\delta} c_m \mathrm{G}_{r+1,3r+1}^{3r,1}\left(D\Big(\frac{x}{\mu_r}\Big) \middle| \begin{matrix} 1, \tau_1 \\ \tau_2, 0 \end{matrix} \right). \tag{7.10}$$

In the above PDF and CDF expressions, $\tau_1 = \Delta(r : \xi^2 + 1)$, $\tau_2 = [\Delta(r : \xi^2), \Delta(r : \alpha), \Delta(r : m)]$ and $\Delta(r : x) = [x/r, \ldots, (x + r - 1)/r]$ [17]. The constants A, B, C, D are defined as $A = \frac{2\alpha^{\alpha/2}}{g^{1+\alpha/2}\Gamma(\alpha)}\left(\frac{g\delta}{g\delta+\Omega'}\right)^{\delta+\{\alpha/2\}}$, $B = \frac{\xi^2\alpha\delta(g+\Omega')}{(\xi^2+1)(g\delta+\Omega')}$, $C = \frac{\xi^2 A}{2^r(2\pi)^{r-1}}$ and $D = B^r/r^{2r}$, whereas $b_m = \binom{\delta-1}{m-1}\frac{(g\delta+\Omega')^{1-m/2}}{(m-1)!}\left(\frac{\Omega'}{g}\right)^{m-1}(\alpha/\delta)^{m/2}\{\alpha\delta/(g\delta+\Omega')\}^{-(\alpha+m)/2}$ and $c_m = b_m r^{\alpha+m-1}$ [17]. As introduced in reference [24], the δ denotes the fading and is a natural number, $g = b_0(1 - \rho)$, where $2b_0$ denotes the total scattering power and g represents the average power of the scattering component. ρ denotes the scattering power coupled with the line-of-sight component, the power of which is denoted by Ω, whereas $\Omega' = \Omega + 2b_0\rho + 2\sqrt{2b_0\rho\Omega}\cos(\phi_A - \phi - B)$ is the average power of the coherent contributions. The deterministic phase terms of line-of-sight and coupled to line-of-sight components have been denoted by ϕ_A and ϕ_B, respectively. The electrical SNR is denoted as μ_r, which can be defined as $\mu_r = \mathbb{E}\left[\frac{|g_{23}|^r P_F}{\sigma_{n2}^2}\right]$. The pointing error parameter is ξ, which is the ratio of the equivalent beamwidth and jitter standard deviation σ_s [26]. For selection of various pointing error parameters and definitions of related variables, readers are suggested to refer [26].

7.4 Performance Analysis

In this section, the key performance metrics such as outage probability and ergodic capacity have been derived.

7.4.1 *Exact Outage Probability*

The reliability of the mixed RF/FSO DF NOMA system can be evaluated through the outage probability, which is defined as the event where the instantaneous end-to-end signal-to-noise ratio (SNR) falls below a predefined threshold SNR of γ_{th}. This threshold is determined based on the quality-of-service (QoS) requirements for the communication system.

Mathematically, the outage probability P_{out,s_1} for a given link for symbol s_1 can be expressed as

$$P_{out,s_1} = \Pr\left[\min(\gamma_{12,s1}, \gamma_{13,s1}) \leq \gamma_{th_1}\right] \tag{7.11}$$

where γ_{th_1} is the threshold SNR for s_1. The statistics of random variable$Y$can be defined as $Y = \min(\gamma_{12,s1}, \gamma_{13,s1}) = (\frac{a_1\lambda_{12}}{a_2\lambda_{12}+1}, \frac{a_1\lambda_{13}}{a_2\lambda_{13}+1})$. The CDF of Y

$$\Pr\left[\frac{a_1\lambda_{1j}}{a_2\lambda_{1j}+1} \leq \gamma\right] = \begin{cases} F_{\lambda_{1j}}\left(\frac{\gamma}{a_1 - a_2\gamma}\right), & \text{if } \gamma < \frac{a_1}{a_2} \\ 1, & \text{otherwise} \end{cases} \tag{7.12}$$

where $j = \{2, 3\}$. The outage probability for symbol s_1, P_{out,s_1} can be computed by replacing γ by γ_{th_1} in $F_Y(\gamma)$ and therefore is given as

$$P_{\text{out},s_1} = \frac{\gamma\left(m_{12}, \beta_{12}\frac{\gamma_{\text{th}_1}}{a_1 - a_2\gamma_{\text{th}_1}}\right)}{\Gamma(m_{12})} + \frac{\gamma\left(m_{13}, \beta_{13}\frac{\gamma_{\text{th}_1}}{a_1 - a_2\gamma_{\text{th}_1}}\right)}{\Gamma(m_{13})} - \frac{\gamma\left(m_{12}, \beta_{12}\frac{\gamma_{\text{th}_1}}{a_1 - a_2\gamma_{\text{th}_1}}\right)\gamma\left(m_{13}, \beta_{13}\frac{\gamma_{\text{th}_1}}{a_1 - a_2\gamma_{\text{th}_1}}\right)}{\Gamma(m_{12})\Gamma(m_{13})} \tag{7.13}$$

The definition of P_{out,s_2} for symbol s_2 can be developed from the fact that for DF relaying protocol, the end-to-end rate is dominated by the weakest link [23]. Therefore, P_{out,s_2} can be defined as

$$P_{\text{out},s_2} = \Pr[X \leq \gamma_{\text{th}_2}], \quad X = \min(a_2\lambda_{12}, \gamma_{23,s_2}) \tag{7.14}$$

After performing some mathematical manipulations, the outage probability corresponding to the transmission of symbol s_2 can be expressed as

$$P_{\text{out},s_2} = \frac{\gamma\left(m_{12}, \beta_{12}\frac{\gamma_{\text{th}_2}}{a_2}\right)}{\Gamma(m_{12})} + C\sum_{m=1}^{\delta} c_m G_{r+1,3r+1}^{3r,1}\left(\frac{D\gamma_{\text{th}_2}}{\mu_r}\middle| \begin{matrix} 1, \tau_1 \\ \tau_2, 0 \end{matrix}\right) - \frac{C}{\Gamma(m_{12})}\sum_{m=1}^{\delta} c_m \times G_{r+1,3r+1}^{3r,1}\left(\frac{D\gamma_{\text{th}_2}}{\mu_r}\middle| \begin{matrix} 1, \tau_1 \\ \tau_2, 0 \end{matrix}\right)\gamma\left(m_{12}, \beta_{12}\frac{\gamma_{\text{th}_2}}{a_2}\right) \tag{7.15}$$

7.4.2 Asymptotic Outage Probability

Although the exact results presented in the previous section are important, they involve computationally extensive mathematical functions. In order to quickly extract insights into the performance of the underlying system, an approximate analysis becomes indispensable. Therefore, a high SNR asymptotic derivation is provided to comprehend the exact analytical results in terms of simple elementary functions. In the high SNR regime, the system's performance can be approximated using simplified expressions, which provide valuable insights without the need for complex computations. These approximations offer a convenient way to understand the behavior of the system under idealized conditions, facilitating rapid evaluation and comparison of different system configurations and parameters. The high SNR asymptotic analysis involves simplifying the expressions for outage probability and ergodic capacity by focusing on the dominant terms as SNR approaches infinity.

By neglecting nondominant terms and applying mathematical approximations, such as Taylor series expansions or asymptotic limits, closed-form expressions can be derived that capture the system's behavior at high SNR levels. These approximate expressions offer a practical way to assess the system's performance in scenarios where high SNR operation is common, such as in line-of-sight communication or when using high-power transmitters. They provide valuable insights into the fundamental limits of the system's performance and can guide the design and optimization of communication systems for efficient operation in high SNR environments. From [27, Eq. (8.11.4)], for small values of z, $\gamma(a, z) \simeq \frac{z^a}{a}$.

$$P_{\text{out},s_1} \simeq \frac{\left(\beta_{12}\frac{\gamma_{\text{th}_1}}{a_1 - a_2\gamma_{\text{th}_1}}\right)^{m_{12}}}{m_{12}\Gamma(m_{12})} + \frac{\left(\beta_{13}\frac{\gamma_{\text{th}_1}}{a_1 - a_2\gamma_{\text{th}_1}}\right)^{m_{13}}}{m_{13}\Gamma(m_{13})} \tag{7.16}$$

If $\tau_m = \min(\tau_2)$, then as $\mu_r \to 0$, the Meijer-G function can be presented as given below using [28, Eq. (07.34.06.0006.01)]

$$P_{\text{out},s_2} \simeq \frac{\left(\beta_{12}\frac{\gamma_{\text{th}_2}}{a_2}\right)^{m_{12}}}{m_{12}\Gamma(m_{12})} + C\sum_{m=1}^{\delta} c_m \left(\frac{D\gamma_{\text{th}_2}}{\mu_r}\right)^{\tau_m} \frac{\prod_{i=1}^{3r}\Gamma(\tau_{2,i} - \tau_m)}{\tau_m \prod_{i=2}^{r}\Gamma(\tau_{1,i} - \tau_m)} \tag{7.17}$$

Comparing the asymptotic expressions presented in (7.16) and (7.17) with those provided in (7.13) and (7.15), it can be easily noted that approximate expressions provide simpler representations. These results can be utilized by the system design engineers to gain insights regarding the significant parameters that impact the end-to-end outage probability. Later, in the result section, these results have been validated against their exact counterparts.

7.5 Ergodic Capacity Analysis

From an information theory perspective, ergodic capacity is a fundamental metric that quantifies the data rate that a communication channel can reliably provide in a noisy environment. It represents the maximum achievable rate of information transmission over the channel, taking into account the channel's characteristics and the presence of noise. In the context of the mixed RF/FSO DF NOMA system, the existence of both a direct link and an indirect link between the source and destination through the relay node enhances the overall data rate. The direct link provides a reliable path for communication, while the relay node facilitates additional data transmission through cooperative relaying, thereby increasing the system's capacity. The ergodic capacity of the system captures the statistical average of the achievable data rates over time, considering the variability of channel conditions and the presence of noise. It provides a comprehensive measure of the system's performance, taking into account both the instantaneous channel conditions and the long-term behavior of the communication channel. By analyzing the ergodic capacity, insights can be gained into the system's ability to support data transmission at varying rates under different channel conditions. This allows for the optimization of communication protocols and resource allocation strategies to maximize the overall data throughput while ensuring reliable communication in the presence of noise and fading. In fact, for symbol s_1, the capacity offered in the DF relay setup is dominated by the weakest link. Therefore, the capacity can be derived as $C_1 = 0.5\mathbb{E}[\min\{\log_2(1+\gamma_{12,s1}), \log_2(1+\gamma_{13,s1})\}]$. If $U = \min\{\lambda_{12}, \lambda_{13}\}$, then the ergodic capacity can be expressed as

$$C_1 = 0.5\mathbb{E}\left[\log_2\left(1+\frac{Ua_1}{a_2U+1}\right)\right] = 0.5\mathbb{E}\left[\log_2\left(1+(a_1+a_2)U\right)-\log_2\left(1+a_2U\right)\right].$$

Since $F_U(x) = F_{\lambda_{12}}(x) + F_{\lambda_{13}}(x) - F_{\lambda_{12}}(x)F_{\lambda_{13}}(x)$, the following integrals can be defined for deriving C_1:

$$\begin{aligned}\mathcal{I}_1(i,k) &= \int_0^\infty \log(1+a_kx)f_{\lambda_{1j}}(x)dx \\ &= \frac{\beta_{1i}^{m_{1i}}}{\Gamma(m_{1i})}\int_0^\infty x^{m_{1i}-1}e^{-\beta_{1i}x}\mathrm{G}_{2,2}^{1,2}\left(a_kx\,\middle|\,\begin{matrix}1,1\\1,0\end{matrix}\right)dx\end{aligned} \tag{7.18}$$

$$\begin{aligned}\mathcal{I}_2(i,j,k) &= \int_0^\infty \log(1+a_kx)F_{\lambda_{1j}}(x)f_{\lambda_{1i}}(x)dx \\ &= \frac{\beta_{1i}^{m_{1i}}}{\Gamma(m_{1i})}\int_0^\infty x^{m_{1i}-1}e^{-\beta_{1i}x}\gamma\left(m_{1j},\beta_{1j}x\right) \\ &\quad\times \mathrm{G}_{2,2}^{1,2}\left(a_kx\,\middle|\,\begin{matrix}1,1\\1,0\end{matrix}\right)dx\end{aligned} \tag{7.19}$$

where $i \in \{2,3\}$, $j \in \{2,3\}$ and $k \in \{2,3\}$. Moreover, with the aid of [28, Eq. (07.34.21.0088.01)], the closed-form solution to $\mathcal{I}_1(i,k)$ can be formulated as

$$\mathcal{I}_1(i,k) = \frac{1}{\Gamma(m_{1i})}\mathrm{G}_{3,2}^{1,3}\left(\frac{a_k}{\beta_{1i}}\,\middle|\,\begin{matrix}1-m_{1i},1,1\\1,0\end{matrix}\right) \tag{7.20}$$

Also, applying [29, Eq. (2.3)], the closed-form expression of $\mathcal{I}_2(i,j,k)$ can be obtained as

$$\mathcal{I}_2(i,j,k) = \frac{1}{\Gamma(m_{1i})\Gamma(m_{1j})}\mathrm{H}\left[\begin{matrix}\begin{pmatrix}0,1\\1,0\end{pmatrix}\\ \begin{pmatrix}1,1\\1,2\end{pmatrix}\\ \begin{pmatrix}1,2\\2,2\end{pmatrix}\end{matrix}\,\middle|\,\begin{matrix}(1-m_{1i}:1,1)\\ \overline{} \\ (1,1)\\ (m_{1,j},1),(0,1)\\ (1,1),(1,1)\\ (1,1),(0,1)\end{matrix}\,\middle|\,\frac{\beta_{1j}}{\beta_{1i}},\frac{a_k}{\beta_{1i}}\right] \tag{7.21}$$

After having formulated both the integrals, expression for C_1 can be expressed as

$$C_1 = \frac{1}{2}[\mathcal{I}_1(2,3) + \mathcal{I}_1(3,3) - \mathcal{I}_2(3,2,3) - \mathcal{I}_2(2,3,3) - \mathcal{I}_1(2,2) - \mathcal{I}_1(3,2) + \mathcal{I}_2(3,2,2) + \mathcal{I}_2(2,3,2)] \tag{7.22}$$

For symbol s_2, the ergodic capacity can be evaluated as

$$C_2 = \frac{1}{2}\mathbb{E}\Big[\min\Big\{\log_2(1+\gamma_{12,s2}), \log_2(1+\gamma_{23,s2})\Big\}\Big] = \frac{1}{2}\mathbb{E}\Big[\log_2\Big(1+\min\{\gamma_{12,s2}, \gamma_{23,s2}\}\Big)\Big] \tag{7.23}$$

Considering $W = \min\{a_2\lambda_{12}, \gamma_{23,s2}\}$, the PDF can be expressed as $f_W(x) = f_{a_2\lambda_{12}}(x) + f_{\gamma_{\text{th}}}(x) - f_{a_2\lambda_{12}}(x)F_{\gamma_{23,s2}}(x) - F_{a_2\lambda_{12}}(x)f_{\gamma_{23,s2}}(x)$. Substituting the expressions for $F_{\gamma_{23,s2}}(x)$, $f_{\gamma_{23,s2}}(x)$, $F_{\lambda_{12}}(x)$, and $f_{\lambda_{12}}(x)$, following integrals are to be addressed in order to derive the closed-form expression for the ergodic capacity

$$\mathcal{I}_3 = \frac{\beta_{12}^{m_{12}}}{a_2{}^{m_{12}-1}\Gamma(m_{12})}C\sum_{m=1}^{\delta}c_m\int_0^\infty x^{m_{12}-1}\exp\Big(-\frac{\beta_{12}}{a_2}x\Big)\mathrm{G}_{2,2}^{1,2}\Big(x\Big|\begin{matrix}1,1\\1,0\end{matrix}\Big) \times \mathrm{G}_{r+1,3r+1}^{3r,1}\Big(D\Big(\frac{x}{\mu_r}\Big)\Big|\begin{matrix}1,\tau_1\\\tau_2,0\end{matrix}\Big)dx \tag{7.24}$$

$$\mathcal{I}_4 = \frac{1}{\Gamma(m_{12})}\frac{\xi^2 A}{2^r}\sum_{m=1}^{\delta}b_m\int_0^\infty\frac{1}{x}\gamma\Big(m_{12},\frac{\beta_{12}}{a_2}x\Big) \times \mathrm{G}_{2,2}^{1,2}\Big(x\Big|\begin{matrix}1,1\\1,0\end{matrix}\Big)\mathrm{G}_{1,3}^{3,0}\Big(B\Big(\frac{x}{\mu_r}\Big)^{1/r}\Big|\begin{matrix}\xi^2+1\\\xi^2,\alpha,m\end{matrix}\Big)dx \tag{7.25}$$

$$\mathcal{I}_5 = \frac{\beta_{12}^{m_{12}}}{a_2{}^{m_{12}-1}\Gamma(m_{12})}\int_0^\infty x^{m_{12}-1}\exp\Big(-\frac{\beta_{12}}{a_2}x\Big)\mathrm{G}_{2,2}^{1,2}\Big(x\Big|\begin{matrix}1,1\\1,0\end{matrix}\Big)dx \tag{7.26}$$

$$\mathcal{I}_6 = \frac{\xi^2 A}{2^r}\sum_{m=1}^{\delta}b_m\int_0^\infty\frac{1}{x}\mathrm{G}_{2,2}^{1,2}\Big(x\Big|\begin{matrix}1,1\\1,0\end{matrix}\Big)\mathrm{G}_{1,3}^{3,0}\Big(B\Big(\frac{x}{\mu_r}\Big)^{1/r}\Big|\begin{matrix}\xi^2+1\\\xi^2,\alpha,m\end{matrix}\Big)dx \tag{7.27}$$

For evaluating integral $\mathcal{I}_3$, the identity [29, Eq. (2.3)] can be utilized with the aid of which a closed-form solution has been expressed in (7.28).

$$\mathcal{I}_3 = \frac{\beta_{12}}{a_2\Gamma(m_{12})}\frac{\xi^2 A}{2^r}\sum_{m=1}^{\delta} b_m \mathrm{H}\left[\begin{pmatrix}0,1\\1,0\end{pmatrix}\begin{pmatrix}1,2\\2,2\end{pmatrix}\begin{pmatrix}3r,1\\r+1,3r+1\end{pmatrix}\left|\begin{matrix}(1-m_{12}:1,1)\\ \overline{}\\ (1,1),(1,1)\\(1,1),(0,1)\\(1,1),(\tau_1,1_{\mathrm{len}\tau_1})\\(\tau_2,1_{\mathrm{len}\tau_2}),(0,1)\end{matrix}\right|\frac{a_2}{\beta_{12}},\frac{Ba_2}{\mu_r\beta_{12}}\right] \tag{7.28}$$

To the best of author's knowledge, no closed-form solution can be obtained for the integral mentioned above in (7.25). However, using [30, Eq. (8.352.1)], the incomplete Gamma function can be represented in series expansion for integer values of m_{12}, to produce a resulting integral as given below

$$\begin{aligned}\mathcal{I}_4 = &\frac{(m_{12}-1)!}{\Gamma(m_{12})}\frac{\xi^2 A}{2^r}\sum_{m=1}^{\delta} b_m\Bigg[\int_0^{\infty}\frac{1}{x}\exp\Big(-\frac{\beta_{12}}{a_2}x\Big)\mathrm{G}_{2,2}^{1,2}\left(x\left|\begin{matrix}1,1\\1,0\end{matrix}\right.\right)\\ &\mathrm{G}_{1,3}^{3,0}\left(B\Big(\frac{x}{\mu_r}\Big)^{1/r}\left|\begin{matrix}\xi^2+1\\ \xi^2,\alpha,m\end{matrix}\right.\right)\\ &-\sum_{i_1=0}^{m_{12}}\Big(\frac{\beta_{12}}{a_2}\Big)^{i_1}\frac{1}{i_1}\int_0^{\infty}x^{i_1-1}\exp\Big(-\frac{\beta_{12}}{a_2}x\Big)\mathrm{G}_{2,2}^{1,2}\left(x\left|\begin{matrix}1,1\\1,0\end{matrix}\right.\right)\\ &\times\mathrm{G}_{1,3}^{3,0}\left(B\Big(\frac{x}{\mu_r}\Big)^{1/r}\left|\begin{matrix}\xi^2+1\\ \xi^2,\alpha,m\end{matrix}\right.\right)\Bigg]\end{aligned} \tag{7.29}$$

$$\begin{aligned}\mathcal{I}_4 = &\frac{(m_{12}-1)!}{\Gamma(m_{12})}C\sum_{m=1}^{\delta} c_m \mathrm{G}_{r+2,3r+2}^{3r+2,1}\left(\frac{B^r}{\mu_r r^{2r}}\left|\begin{matrix}0,\tau_1,1\\ \tau_2,0,0\end{matrix}\right.\right)\\ &-\frac{(m_{12}-1)!}{\Gamma(m_{12})}C\sum_{m=1}^{\delta} b_m\sum_{i_1=0}^{m_{12}-1}\frac{1}{(i_1)!}\\ &\times\mathrm{H}\left[\begin{pmatrix}0,1\\1,0\end{pmatrix}\begin{pmatrix}1,2\\2,2\end{pmatrix}\begin{pmatrix}3,0\\1,3\end{pmatrix}\left|\begin{matrix}(1-\frac{i_1}{r}/r:1,\frac{1}{r})\\ \overline{}\\ (1,1),(1,1)\\(1,1),(0,1)\\(\xi^2+1,1)\\(\xi^2,1),(\alpha,1),(m,1)\end{matrix}\right|\frac{a_2}{\beta_{12}},\frac{Ba_2^{\frac{1}{r}}}{(\beta_{12}\mu_r)^{\frac{1}{r}}}\right]\end{aligned} \tag{7.30}$$

The integrals $\mathcal{I}_5$ and $\mathcal{I}_6$ can be solved using [28, Eq. (07.34.21.0088.01)] and [28, Eq. (07.34.21.0013.01)], respectively, to give

$$\mathcal{I}_5 = \frac{a_2}{m_{12}}\mathrm{G}_{3,2}^{1,3}\left(\frac{a_2}{\beta_{12}}\left|\begin{matrix}-m_{12},1,1\\1,0\end{matrix}\right.\right) \tag{7.31}$$

and

$$\mathcal{I}_6 = C\sum_{m=1}^{\delta} c_m \mathrm{G}_{r+2,3r+2}^{3r+2,1}\left(\frac{D}{\mu_r}\middle|\begin{matrix}1,\tau_1,1\\ \tau_2,0,0\end{matrix}\right) \tag{7.32}$$

Finally, the result for C_2 can be expressed as $C_2 = (\log(2))^{-1}(\mathcal{I}_5 + \mathcal{I}_6 - \mathcal{I}_3 - \mathcal{I}_4)$. Of course, the result for ergodic capacity depends not only on the power allocation coefficients but also on the FSO link parameters such as type of optical demodulation, pointing error, strength of atmospheric turbulence apart from fading on both the RF links. More insights are presented about the derived analytical results in the next section.

7.6 Results

The plot in Fig. 7.1 depicts the outage probability curve for the NOMA system for both the symbols. The figure also shows the asymptotic analysis and Monte Carlo simulation results. As the average SNR increases on various links, the outage probability reduces. Also, the asymptotic analysis merges with the exact results for high SNR values. The outage probability has been plotted versus average SNRs on $S - R$, $R - D$, and the direct link $S - D$. The symbol s_1 is sent directly to the destination, and if such a condition prevails, it can be seen in the plot that the outage probability is higher than other conditions. Now, since the symbol s_2 includes transmission through the FSO link, there is a possibility on the dependency of outage on the type of optical demodulation schemes. As $r = 1$ represents the coherent

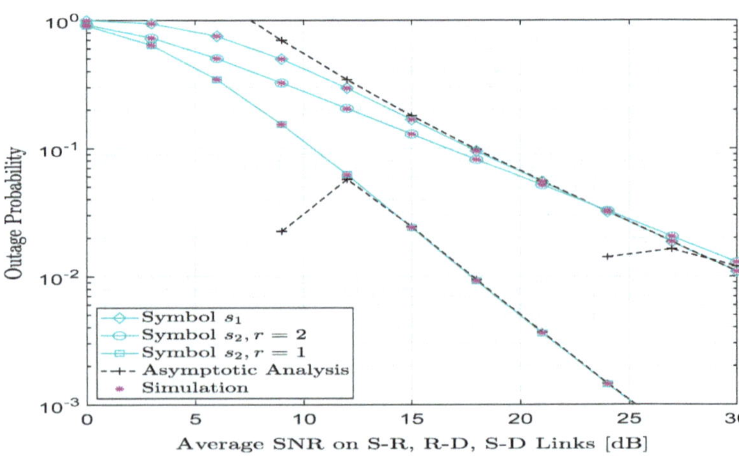

Fig. 7.1 Outage probability analysis for reception of both symbols

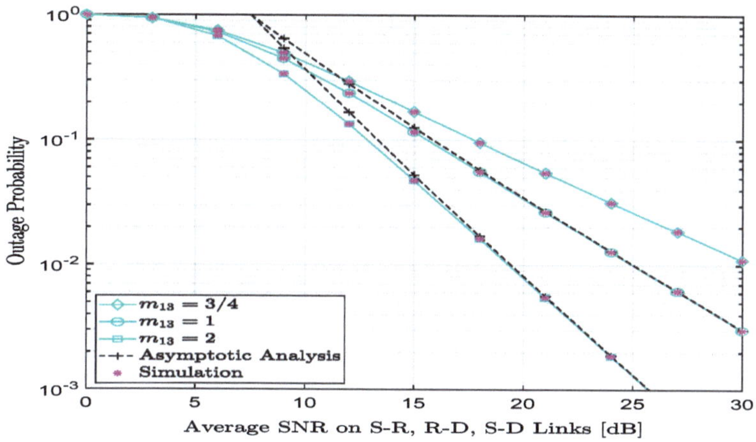

Fig. 7.2 Outage probability analysis for fading strength on the direct link

demodulation, the outage probability of receiving the symbol s_2 is minimum for this case due to the robustness of the demodulation schemes. However, in some situations, it is not possible to maintain the carrier synchronization that is essential for the coherent demodulation scheme, and the intensity based demodulation is required. For this, the variable $r = 2$ is considered, and the outage probability is more in this case, as compared to the case of coherent demodulation.

The outage probability has been demonstrated in plot Fig. 7.2 for recovering the symbol s_1 when the fading strength on the direct link changes. Specifically, the strength of fading on the direct link can be changed using the modified values of the Nakagami-m parameter of m_{13}. As the values of the fading parameter increase, the multipath conditions become less severe. This enhances the reliability of the network. For the purpose of the study, we have considered $m_{13} = [3/4, 1, 2]$. It can be seen in the plot as the value of m_{13} increases, the outage probability decreases, suggesting better channel conditions. The plot also shows the asymptotic high SNR analysis that matches with the exact results for high SNR.

On the other hand, the outage probability also depends on the characteristics of the FSO link. For the study, the pointing error between the optical transmitter and receiver has been varied, and the outage probability of receiving symbol s_2 has been recorded. Apart from this, the impact of the optical demodulation scheme has also been depicted in Fig. 7.3. The degradation due to the misalignment between the transmitter and the receiver is considered using the pointing error parameter $\xi = [0.7, 2]$. Importantly, the smaller value of the pointing error parameter shows more misalignment error. This combined with the IM/DD demodulation scheme shows the worst performance. The probability decreases, even with the IM/DD demodulation scheme, when the pointing error is minimized. This condition corresponds to the combination $\xi = 2, r = 2$. The case when pointing error is large ($\xi = 0.7$) but demodulation is done using the coherent scheme shows better

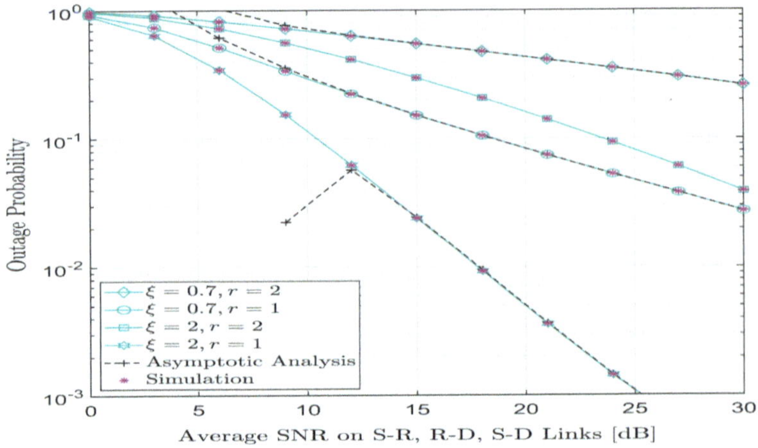

Fig. 7.3 Outage probability analysis for variation of pointing error on the FSO link

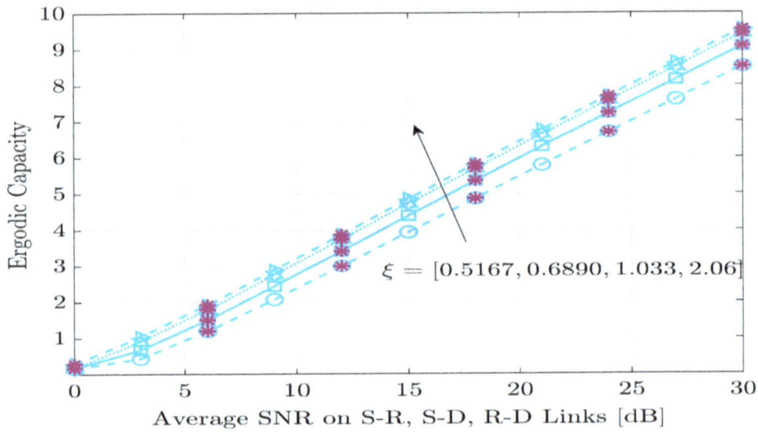

Fig. 7.4 Ergodic capacity analysis for variation of pointing error on the FSO link

performance compared to the previous case. The best case is when $\xi = 2, r = 1$ which shows the lowest outage probability.

The ergodic capacity of the NOMA system has been shown in Fig. 7.4. The impact of pointing error on the optical wireless channel has been analyzed. The fading on the RF links has been represented using the Nakagami-m parameter values of $m_{12}, m_{13} = [3/4, 2]$. The average SNRs on all the three links have been assumed to be equal during simulation. As per the nature of the capacity, the plot indicates that as the average SNRs on various links increase, the throughput also increases. Nevertheless, it is depicted in the plot that as the pointing error is reduced (larger values of the pointing error parameter ξ), the ergodic capacity is enhanced.

Therefore, the effect of pointing error on the FSO link plays an important role in determining the capacity of the system.

7.7 Conclusion

In this chapter, performance analysis of mixed RF/FSO DF NOMA relaying system has been presented. Apart from the blind relay path between the source and destination through relay node, a direct link has also been considered in the analysis. The fading on the RF and FSO links has been modeled using Nakagami-m and Malaga distributions, respectively. The modeling on the FSO link has been performance by considering the impact of pointing error and type of optical demodulation scheme. For NOMA DF system, closed-form expressions for the outage probability and ergodic capacity have been derived. Additionally, the outage results have been extended to present high SNR approximate expressions. The analytical framework has been verified using numerical simulations. It has been established that the various parameters on the FSO link such as pointing error, atmospheric turbulence, and type of optical demodulation along with the strength of RF fading impact the performance of the system. More importantly, the results suggest that the percentage of power allocation assigned to RF and/or FSO link governs the quality of service experienced by the end user.

References

1. H. Dahrouj, A. Douik, F. Rayal, T.Y. Al-Naffouri, M.S. Alouini, Cost-effective hybrid RF/FSO backhaul solution for next generation wireless systems. IEEE Wirel. Commun. **22**(5), 98–104 (2015)
2. E. Lee, J. Park, D. Han, G. Yoon, Performance analysis of the asymmetric dual-hop relay transmission with mixed RF/FSO links. IEEE Photon. Technol. Lett. **23**(21), 1642–1644 (2011)
3. J. Choi, Non-orthogonal multiple access in downlink coordinated two-point systems. IEEE Commun. Lett. **18**(2), 313–316 (2014)
4. Z. Ding, M. Peng, H.V. Poor, Cooperative non-orthogonal multiple access in 5G systems. IEEE Commun. Lett. **19**(8), 1462–1465 (2015)
5. H. Tabassum, E. Hossain, J. Hossain, Modeling and analysis of uplink non-orthogonal multiple access in large-scale cellular networks using Poisson cluster processes. IEEE Trans. Commun. **65**(8), 3555–3570 (2017)
6. Z. Zhang, Z. Ma, M. Xiao, Z. Ding, P. Fan, Full-duplex device-to-device-aided cooperative nonorthogonal multiple access. IEEE Trans. Veh. Technol. **66**(5), 4467–4471 (2017)
7. X. Yue, Y. Liu, S. Kang, A. Nallanathan, Z. Ding, Exploiting full/half-duplex user relaying in NOMA systems. IEEE Trans. Commun. **66**(2), 560–575 (2018)
8. T.N. Do, D.B. da Costa, T.Q. Duong, B. An, Improving the performance of cell-edge users in NOMA systems using cooperative relaying. IEEE Trans. Commun. **66**(5), 1883–1901 (2018)
9. M. Najafi, V. Jamali, P.D. Diamantoulakis, G.K. Karagiannidis, R. Schober, Non-orthogonal multiple access for FSO backhauling, in *2018 IEEE Wireless Communications and Networking Conference (WCNC)* (2018), pp. 1–6

10. M.V. Jamali, S.M. Azimi-Abarghouyi, H. Mahdavifar, *Outage Probability Analysis of Uplink NOMA over Ultra-High-Speed FSO-Backhauled Systems* (2018), pp. 1–6
11. P. Saxena, M.R. Bhatnagar, 1-bit feedback-based beamforming scheme for an uplink FSO-NOMA system with SIC errors. Appl. Opt. **59**(36), 11274–11291 (2020)
12. M.V. Jamali, H. Mahdavifar, Uplink non-orthogonal multiple access over mixed RF-FSO systems. IEEE Trans. Wirel. Commun. **19**(5), 3558–3574 (2020)
13. A. Upadhya, V.K. Dwivedi, G. Singh, Relay-aided free-space optical communications using $\alpha - \mu$ distribution over atmospheric turbulence channels with misalignment errors. Opt. Commun. **416**, 117–124 (2018)
14. A. Upadhya, V.K. Dwivedi, G. Singh, Multiuser diversity for mixed RF/FSO cooperative relaying in the presence of interference. Opt. Commun. **442**, 77–83 (2019)
15. A. Upadhya, V.K. Dwivedi, M.S. Alouini, Interference-limited mixed MUD-RF/FSO two-way cooperative networks over double generalized gamma turbulence channels. IEEE Commun. Lett. **442**, 1551–1555 (2019)
16. V.K. Tonk, A. Upadhya, P.K. Yadav, V.K. Dwivedi, Mixed mud-RF/FSO two way dcode and forward relaying networks in the presence of co-channel interference. Opt. Commun. **464**, 125415 (2020)
17. I.S. Ansari, F. Yilmaz, M.S. Alouini, Performance analysis of free-space optical links over $\mathcal{M}$´*a*laga ($\mathcal{M}$) turbulence channels with pointing errors. IEEE Trans. Wirel. Commun. **15**(1), 91–102 (2016)
18. A. Upadhya, V.K. Dwivedi, G.K. Karagiannidis, On the effect of interference and misalignment error in mixed RF/FSO systems over generalized fading channels. IEEE Trans. Commun. **68**(6), 3681–3695 (2020)
19. H. AlQuwaiee, I.S. Ansari, M.S. Alouini, On the performance of free-space optical communication systems over double generalized gamma channel. IEEE J. on Selec. Areas Commun. **33**(9), 1829–1840 (2015)
20. A. Goel, A. Upadhya, V.K. Dwivedi, Diversity aided millimeter-wave/free space optical cooperative relaying systems. Int. J. Commun. Syst. **34**(4), e4700 (2021). e4700 dac.4700
21. V.K. Tonk, V.K. Dwivedi, A. Upadhya, S.P. Singh, P.K. Yadav, New results in secrecy analysis of mixed RF/FSO cooperative relaying networks. Opt. Commun. **503**, 127376 (2022)
22. A. Upadhya, J. Gupta, V.K. Dwivedi, M.-S. Alouini, Impact of RF I/Q imbalance on interference-limited mixed RF/FSO TWR systems with non-zero boresight error. IEEE Wireless Commun. Lett. **10**(2), 416–420 (2021)
23. Y.-W. Peter Hong, W.-J. Huang, C.-C. Jay Kuo, *Cooperative Communications and Networking: Technologies and System Design* (Springer, New York, 2010)
24. A.J. Navas, J.M.G. Balsells, J.F. Paris, A.P. Notario, A unifying statistical model for atmospheric optical scintillation, in *Numerical Simulations of Physical and Engineering Processes, Ch. 8* (2011)
25. A. Upadhya, Investigation of mixed RF/FSO decode-and-forward NOMA cooperative relaying networks. Wireless Pers Commun. **124**, 2923–2938 (2022)
26. A.A. Farid, S. Hranilovic, Outage capacity optimization for free-space optical links with pointing errors. J. Lightwave Technol. **25**, 1702–1710 (2007)
27. F. W. J. Olver, D. W. Lozier, R. F. Boisvert, C. W. Clark. 1st ed. New York, NY, USA: Cambridge Univ., Press NIST Handbook of Mathematical Functions (2010)
28. I. Wolfram, Wolfram, research, mathematica edition: Version 10.0. Champaign. (Wolfram Research, Inc., New York, 2010)
29. P. Mittal, K. Gupta, An integral involving generalized function of two variables. Proc. Indian Acad. Sci. Section A **75**(3), 117–123 (1972)
30. I.S. Gradshteyn, I.M. Ryzhik, *Table of Integrals, Series and Products* (Academic, New York, 2000)

Chapter 8
Role of Artificial Intelligence/Machine Learning in Free Space Optical Communication Networks

8.1 Introduction

Previous chapters presented a mathematical modeling-based approach on how to analyze FSO communication networks. However, due to transmission of optical signal through free space, the complex variations in signal make the analysis much more complicated. The role of machine learning/ deep learning models becomes crucial in making an overall improvement in FSO system performances. This chapter provides an overview of how the machine learning models can be used for future wireless communication systems.

8.2 Introduction to Machine Learning Models

The role of computing machines is to take a certain set of inputs and based on the algorithm (function) provided into the system and calculate the outputs of the system. However, with the advent of machine learning algorithms, it is possible to provide the set of input and output values and ask the machine to find out the function that maps the input and output. Machine learning and deep learning are revolutionizing the field of artificial intelligence leading to significant advancements across various domains over the past decade. Machine learning is a subset of artificial intelligence that enables systems to learn and make decisions from data without the need to be programmed. Deep learning, a subset of machine learning, utilizes neural networks with many layers to model and understand complex patterns in large datasets. The autonomous nature of deep learning models makes it easy to consider all aspects of input-output relationship.

At its core, machine learning involves the development of algorithms that can process input data, learn from it, and predict outcomes or make decisions based on that learning. Data collection, data preprocessing, model training, model

A. Upadhya et al., *Free-Space Optical Communication Systems for Next Generation Networks*, https://doi.org/10.1007/978-3-031-74805-9_8

evaluation, and deployment process are typically the important processes involved at several stages during training a machine learning program. The machine learning algorithms can be broadly categorized into unsupervised learning, supervised learning, semi-supervised learning, and reinforcement learning. Supervised learning is the most common form of machine learning. This consists of a labeled dataset where the input data are provided with corresponding output labels. Support vector machines, linear regression, logistic regression, and neural networks are few important algorithms used in supervised learning. Using a loss function, these models learn to map input data to the correct output by minimizing the error in their predictions.

Unsupervised learning, in contrast, deals with unlabeled data. The main goal in unsupervised learning models is to identify underlying patterns or structures within the provided data. Due to large order/dimension of data, clustering and dimensionality reduction are typical tasks in unsupervised learning. The main example of unsupervised learning models is the classification problems. Algorithms like K-means clustering, hierarchical clustering, and principal component analysis (PCA) are widely used. Semi-supervised learning combines properties of both supervised and unsupervised learning models. It leverages a small amount of labeled data along with a large amount of unlabeled data to improve learning accuracy. This approach is particularly useful when labeling data is time-consuming or expensive.

Yet another vital area of machine learning is the reinforcement learning, where an agent learns to make decisions by performing actions in an environment to maximize cumulative reward. Such machine learning models are generally used in the fields of robotics, gaming, and autonomous vehicles. Algorithms such as Q-learning and deep Q-networks (DQNs) are prominent in this field. Deep learning, a more recent advancement, has gained immense popularity due to its success in handling unstructured data like images, audio, and text. Actually, deep learning models are an extension to the classical artificial neural networks (ANNs) where the number of hidden layers is generally much more. Due to a large number of intermediate layers between the input and output layers, these models become capable of understanding the feature and label relation. Deep learning models, or deep neural networks, consist of multiple layers of interconnected neurons, mimicking the human brain's structure. The layers learn to extract increasingly abstract features from the data, enabling the model to understand complex representations.

Apart from conventional deep learning model, convolutional neural networks (CNNs) and recurrent neural networks (RNNs) are two primary architectures in deep learning. CNNs are particularly effective for image and video processing tasks, while RNNs excel in sequence data like time series analysis and natural language processing (NLP). The state-of-the-art results in NLP tasks have been achieved using the transformer models. This has been a realization only due to their ability to handle long-range dependencies and parallelize training. All these models have made tremendous enhancement in computing machine roles to analyze and represent complex engineering roles. Some of the machine learning models that can be implemented for solving various issues in FSO systems have been shown in Fig. 8.1. The success of machine learning and deep learning is driven by the

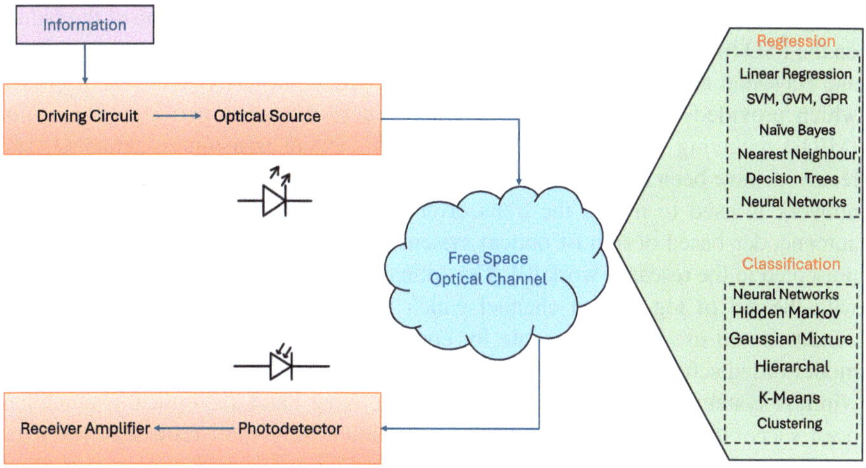

Fig. 8.1 Outage probability analysis for reception of both symbols

availability of large datasets, increased computational power, and advancements in algorithmic research. These technologies have enabled breakthroughs in fields such as healthcare, finance, transportation, and entertainment, transforming industries and improving the quality of life.

8.3 Literature on Machine Learning for FSO Systems

Machine learning has been recently deployed for improvement of operating characteristics of FSO systems. With the help of support vector machines (SVMs), the research work in [1] provides a new method to tackle scintillation in FSO channel using a novel detection scheme. On the forefront of classification problem for channel impairments, various machine learning algorithms have been compared for FSO systems in the work [2]. The authors in [3] have discussed in detail the various machine learning models and their usage in the nonlinear modeling of the FSO systems. The chapter presented the results for estimation of SNR in optical systems using various supervised learning methods such as ANN and SVM. A machine learning-based detection technique is presented in the research work [4] where evaluation is performed based on the input power and the corresponding symbol error rate. A dual CNN-based architecture has been analyzed in the research work [5] where image-based signal processing is implemented to reduce the BER of the FSO system. One of the two neural networks is used to detect the bit and the other to decode it correctly, and the composite system model shows better performance compared to existing optical methods. On the other hand, the authors in [6] implement a gated recurrent unit (GRU) neural network for prediction of FSO fading for mixed RF/FSO relaying system with a far superior performance compared to existing methods of prediction. The statistical distribution of the fading on the FSO link has been compared for GAN, bidirectional long short-term memory and Bayesian neural network in [7], where the proficiency of GANs was discussed which provided relatively less error compared to the other two algorithms. For IM/DD receiving optical system, the chain of optical transmitters, channel, and receivers have been implemented using neural network in [8], where a single neural network is used to model the transceiver system with high accuracy. Moreover, autoencoder-based design of optical system for optical wireless systems has been presented in the research work [9]. The optical channel statistics have been modeled with the aid of log-normal channel which is used to represent weak turbulence condition and use block error rate for comparison of the autoencoder model with model-based schemes. Similarly, autoencoders have been used to model the optical wireless systems in research work [10]. On the other hand, the optical channel has been successfully modeled using the bidirectional long short-term memory deep learning model in [11]. The results presented in [12] provide the GAN based channel modeling for the optical systems which can learn the channel transfer function accurately.

8.4 Machine Learning Models in Wireless Systems

Several key tasks have been accomplished in the past using the concept of machine learning models. The major reason behind the advent and application of machine learning models is the requirement for autonomous systems. One such application is modulation classification in wireless systems. On the physical layer, it is important to decode the modulation scheme that has been used while sending the data from the transmitter. The ability of the machine learning models to classify among the set of data is used by the authors of [13] for modulation classification. This scheme utilized the CNN model for automatic modulation classification (AMC) which proved to be more efficient than the conventional feature-based or maximum likelihood (ML)-based approach. The modulation classification based on data augmentation is studied using deep learning network in the work [14]. The data augmentation methods applied in the investigation have shown capability to handle large volume of data. The modulation classification has been performed using the (CNN)-based cooperative AMC (Co-AMC) method for the MIMO systems in the research work [15]. A method for modulation classification based on image processing has been presented by the authors in [16]. The signal-based CNN (SBCNN) was used for the modulation classification.

The authors in [17] propose a method for the blind identification of Low-Density Parity-Check (LDPC) codes using deep learning techniques. The research introduces a deep learning model that can identify the structure of LDPC codes without requiring prior knowledge about the code parameters. This method significantly enhances the ability to detect and decode LDPC codes in communication systems where the code parameters are not known in advance. The research work

[18] explores the application of deep learning to decode 5G New Radio (5G-NR) communication signals. The authors develop a deep learning-based decoder that can effectively handle the complexity and variability of 5G-NR signals. The proposed method demonstrates improved decoding performance over traditional algorithms, showcasing the potential of deep learning in enhancing the efficiency and reliability of 5G communication systems. The authors of [19] present a unified decoder that leverages deep learning to decode both polar and LDPC codes, which are essential for 5G communication systems. The proposed deep learning-based decoder is designed to optimize the decoding process for these two types of codes, demonstrating significant improvements in decoding accuracy and computational efficiency compared to traditional methods. This unified approach simplifies the implementation of decoders in 5G systems. The research paper published in [20] introduces a low-complexity belief propagation-based decoding scheme for polar codes. The proposed method includes mechanisms for decodability detection and early stopping prediction, which enhance the decoding process by reducing unnecessary computations. This results in a more efficient decoding algorithm with reduced complexity, making it suitable for practical implementations in communication systems.

Various other works have also been reported for security enhancement, channel estimation, and resource allocation in the past. Security of IoT devices with respect to over-the-air attacks has been studied using adversarial machine learning models in [21]. A deep learning model has been used to classify between the intended and the intruder message, and using the detection it is activated to modify the transmitted signal to fool the intrusion. Security issues in 6G systems have been addressed in the research work [22]. A dynamic defense mechanism has been studied where deep learning is used to minimize the effect of security breach in 6G systems. Moreover, the work in [23] studies the security aspect of Wireless Local Area Network (WLAN). A conditional deep belief network is used to detect potential threats in communication system with higher accuracy. Furthermore, cyberattack minimization using deep learning systems has been advocated in the work [24]. The work intends to deal with vulnerability detection, intrusion detection, and fault diagnosis using autonomous deep learning models to eliminate the manual tracking of security. In yet another significant contribution, the authors of [25] analyze security issues with caching IoT systems by tracking the data traffic.

From the perspective of resource allocation, traditional methods are being widely substituted using the deep learning methods. It is important to deliberate that with an efficient resource allocation scheme, the minimization of interference, enhancing the performance and increasing the network efficiency, is possible. However, only mathematical modeling approach was used initially to provide a solution to the resource allocation problem in the wireless communication systems. A research work that aimed at substituting the mathematical modeling for resource allocation with deep learning models has been published in [26]. An ad hoc networking scenario has been considered to use the deep reinforcement learning for the purpose of efficient resource allocation. In another work, the authors of [27] implement a supervised deep neural network for the purpose of allocation of sub-band and power in a multicell environment. From the perspective of CSI sharing, the research work [28] trains a combination of both supervised and unsupervised deep neural networks to establish an efficient resource allocation scheme. The problem of resource allocation in diverse QoS 5G wireless systems has been dealt in work [29]. A cascaded set of two neural networks is utilized to allocate bandwidth and power to users for enhancing the user experience. The proposed system outperforms the scheme where fully connected neural networks are employed for resource allocation. The authors in [30] implement chain of deep neural networks to efficiently allocate the resources for TV services in 5G systems. The work makes use of LSTM model to track the traffic and then implements deep reinforcement learning to estimate the network traffic. The authors of [31] have also considered the power of deep neural network for resource allocation in wireless communication systems.

8.5 Scope of Machine Learning in FSO System

Various aspects of wireless communication systems have been addressed in the past with the help of deep learning. However, not much has been worked upon in the context of optical wireless systems. Due to high bandwidth and high switching speeds, FSO systems can behave much differently to conventional RF systems. Nevertheless, the automatic solutions that can be realized with the help of deep learning models can be promising for the improvement of such systems. The complex nature of FSO systems needs to be modeled effectively before their physical implementation is done. Some of the future research works that can be opted for implementing deep learning in FSO systems are discussed in the section. The following section provides an overview of how machine learning/deep learning can be utilized in optical wireless systems. Various problems that need to be solved efficiently using machine learning models have been illustrated in Fig. 8.2.

8.5.1 Modulation Classification

Modulation recognition is the process of identifying the modulation schemes of received noisy signals. This is crucial for enabling communication between different systems and for military purposes such as monitoring and disrupting enemy communications. Research in modulation recognition has been ongoing for many years, employing traditional algorithms that fall into two main categories: decision-theoretic and pattern recognition methods. The noisy version of signal received is modified in its nature, and various schemes have been employed conventionally to recognize the modulation schemes in the past [32, 33]. A lot of work has been published till date in this domain on the context of RF channels. Nevertheless, the requirements of modulation schemes in FSO systems are different

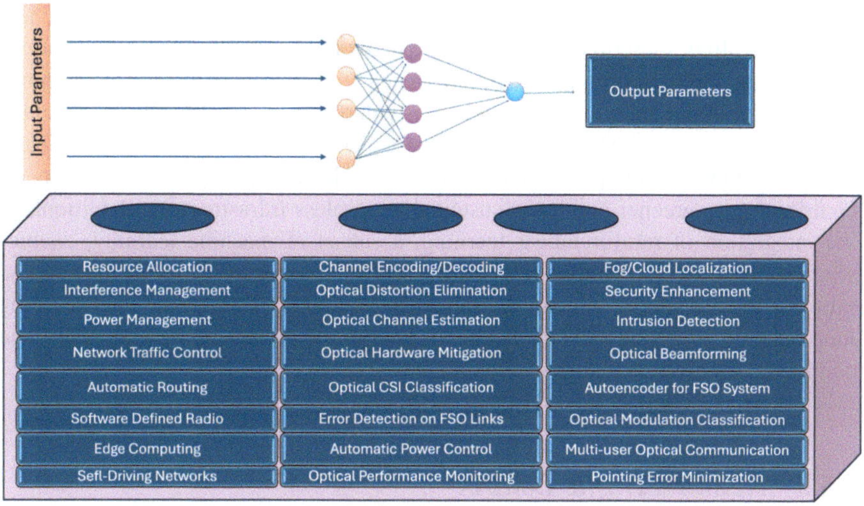

Fig. 8.2 Outage probability analysis for reception of both symbols

and sometimes more stringent. Due to this, there is a requirement to introduce machine learning models for modulation classification in FSO networks. As per the pattern recognition capability of unsupervised learning models, the constellation diagrams of the noisy estimate of the received signal can train a computer vision model to distinguish between several modulation schemes. Based on the position of the received symbol and its nearness to the original transmitted symbol's position, the modulation type can be recognized by the deep learning models. Given the complexity of modulation schemes employed in FSO systems, the deep learning models can be a grate boost in solving this problem.

8.5.2 Resource Allocation

Resource allocation in wireless systems is critical for optimizing performance, ensuring efficient usage of available bandwidth, and maintaining the QoS for users. The resource allocation has conventionally been dealt with using mathematical modeling approach. The advent of machine learning models has revolutionized this aspect of wireless communications. The same needs to be implemented in the context of FSO systems. Machine learning algorithms can process vast amounts of data from network traffic, user behavior, and environmental conditions to estimate demand and allocate resources dynamically. This predictive capability allows for more efficient spectrum utilization, reducing congestion and enhancing overall network throughput. Machine learning models can adapt to changing network conditions in real time, enabling the wireless system to balance loads, avoid interference, and prioritize critical communications effectively. For example, in

5G networks, where the density of connected devices is unprecedented, machine learning facilitates intelligent slicing of the network to cater to diverse applications with varying QoS requirements, from low-latency communications for autonomous vehicles to high-bandwidth demands for video streaming. Moreover, machine learning-driven resource allocation increases energy efficiency by dynamically adjusting transmission power and scheduling sleep cycles for network nodes, contributing to greener and more sustainable wireless infrastructure. Additionally, the integration of reinforcement learning, a subset of machine learning, enables wireless systems to continuously learn from their operational environment and improve their resource allocation strategies over time. This continuous learning loop helps in addressing the challenges posed by the increasing heterogeneity and complexity of modern wireless networks.

8.5.3 Network Traffic Control

Network traffic control in optical wireless systems is pivotal for maximizing performance, ensuring reliability, and maintaining seamless communication. In the upcoming 6G systems, the number of devices that will be connected with each other will be very large. A lot of traffic management is required. The integration of machine learning models into these systems significantly enhances their capability to manage and optimize traffic effectively. Optical wireless communication, with its high bandwidth and rapid data transfer rates, faces challenges such as interference, dynamic environmental conditions, and varying user demands. Machine learning addresses these challenges by providing intelligent traffic prediction and adaptive control mechanisms. Machine learning models can analyze historical and real-time data to predict traffic patterns and potential bottlenecks. By understanding these patterns, the system can preemptively allocate resources, thereby avoiding congestion and ensuring smooth data flow. For example, during peak usage times, machine learning algorithms can dynamically adjust the allocation of bandwidth and prioritize critical data, maintaining QoS for all users. This predictive capability is crucial in environments where high-speed data transmission and low latency are essential, such as in smart cities, autonomous vehicles, and augmented reality applications.

Moreover, machine learning facilitates the efficient management of the optical spectrum. It can detect and mitigate interference, optimize signal strength, and adjust transmission parameters in real time. Techniques such as reinforcement learning enable the system to continuously learn and adapt its traffic control strategies based on the operational environment, leading to ongoing improvements in network performance. Additionally, machine learning-driven traffic control supports energy efficiency. By intelligently managing the traffic load and adjusting transmission power, these systems can reduce energy consumption, contributing to more sustainable network operations.

8.5.4 Edge Computing

Edge computing plays a crucial role in the evolution of optical wireless systems, particularly when combined with machine learning models. Optical wireless systems, known for their high bandwidth and fast data transmission, can benefit immensely from edge computing, which brings data processing closer to the data source. This can reduce latency, enhance real-time data processing capabilities, and can improve overall system efficiency. In the context of optical wireless systems, edge computing can enable rapid and intelligent decision-making, essential for applications demanding low latency and high reliability, such as autonomous vehicles, smart cities, and augmented reality. Machine learning models integrated with edge computing further amplify these benefits. Machine learning algorithms can analyze a large amount of data locally at the edge, providing immediate insights and enabling dynamic resource allocation, interference management, and traffic control. For upcoming 6G systems, this joint venture of machine learning with edge computing will be the key to enhance the user experience as per the QoS needs. This real-time processing capability ensures that the optical wireless network can adapt swiftly to changing conditions and user demands. For instance, in a smart city scenario, edge-based machine learning models can analyze traffic data to optimize communication channels, reduce congestion, and improve the quality of service for critical applications like emergency response systems.

8.5.5 Channel Encoding and Decoding

Channel encoding and decoding are fundamental processes in optical wireless systems, crucial for ensuring reliable data transmission over noisy and error-prone communication channels. The utilization of machine learning models in these processes offers significant advancements, enhancing the system's ability to cope with various transmission challenges and improving overall performance. The main objective of the channel encoding and decoding is to reduce the impact of noise in the channel. In optical wireless systems, channel encoding involves transforming the data into a format that can withstand transmission errors, while decoding reconstructs the original data at the receiver end. This can be done before the electrical signal modulates the optical source such as an LSD or the LASER. Traditional methods rely on predefined algorithms that may not adapt well to dynamic and complex environments. However, machine learning models bring adaptability and intelligence to these tasks, learning from the data and continuously optimizing the encoding and decoding processes.

Machine learning algorithms can analyze large datasets to identify patterns and correlations that are not apparent through conventional methods. This capability allows machine learning models to develop more efficient and robust encoding schemes that maximize data integrity and minimize the error rate. Implementation of machine learning models for channel encoding can be much more efficient. For instance, neural networks can be trained to predict and correct errors in real time, significantly enhancing the reliability of optical wireless communications. Moreover, machine learning models can adapt to changing channel conditions. In optical wireless systems, factors like atmospheric turbulence, physical obstructions, and varying light conditions can affect signal quality. Machine learning algorithms can dynamically adjust encoding parameters based on real-time feedback from the channel, ensuring optimal performance under different scenarios. This adaptability is particularly valuable in environments with fluctuating conditions, such as outdoor optical wireless networks. The implementation of machine learning in channel decoding also improves error correction capabilities. By leveraging techniques like deep learning, the system can better identify and rectify errors introduced during transmission, leading to more accurate data recovery and improved quality of service.

8.5.6 Security Enhancement

Security enhancement in FSO systems is paramount to protect sensitive data and ensure the integrity and confidentiality of communication channels. Although the optical beam is highly directional, still there is the requirement for enhancing the security of FSO systems. The incorporation of deep learning models into these systems significantly boosts their security capabilities, addressing the sophisticated threats and vulnerabilities that modern networks face. Optical wireless systems, known for their high-speed and high-capacity data transmission, are increasingly deployed in critical applications such as finance, healthcare, and military operations, where data security is nonnegotiable. Deep learning models excel in identifying and mitigating security threats due to their ability to analyze vast amounts of data and learn complex patterns. These models can detect anomalies and potential intrusions in real time by continuously monitoring the network traffic and comparing it against known attack patterns. For instance, CNNs and RNNs can be trained to recognize subtle deviations indicative of cyberattacks, such as distributed denial-of-service (DDoS) attacks, eavesdropping, and man-in-the-middle attacks. Various other adversarial deep learning models can be applied for threat detection in optical wireless systems. Once such systems are well trained, they can easily classify and identify the intended user.

One significant advantage of deep learning in security enhancement is its capacity for adaptive learning. Unlike traditional security mechanisms, which rely on predefined rules and signatures, deep learning models can evolve by learning from new data and emerging threats. This continuous learning capability allows optical wireless systems to stay ahead of sophisticated attackers who constantly develop new techniques to breach defenses. Also, deep learning models can enhance encryption methods, making them more robust against attacks. By optimizing cryptographic algorithms and dynamically adjusting encryption keys, these models can provide stronger protection for data in transit. This is particularly important in optical wireless communications, where the high data rates and large volumes of transmitted information present attractive targets for cybercriminals. The application of various deep learning models in the realm of security enhancement will be the key to its usage in high confidential communication scenarios. A lot of work is still to be done before practical implementation of FSO systems in the context of increasing the security.

8.5.7 Interference Management

There are several sources of interference in optical systems. A proper methodology needs to be applied for interference minimization in FSO systems. Traditional methods generally struggle to adapt to the complex and dynamic nature of modern wireless environments. Works like [34] have provided deep reinforcement learning algorithms for interference management. In FSO systems, various sources of in-band emission take place in the vicinity of devices operating on optical frequency range. It is required to enhance the performance of system by detecting such devices and regulate the transmission of optical systems such that the effect of interference is less. For this, the FSO telescope's orientation can be modified such that the impact of interference is less. This rotation of the optical beam can be done by deep learning models. Deep learning models can be trained to detect, classify, and mitigate various types of interference in real time. By analyzing large datasets of wireless signals, these models can identify patterns and anomalies indicative of interference, even in highly congested or noisy environments. Once trained, deep learning models can predict interference scenarios and suggest optimal strategies for resource allocation, beamforming, and power control to minimize the impact of interference. They can dynamically adjust these strategies as network conditions change, maintaining optimal performance without human intervention.

8.5.8 Channel Estimation

Conventionally, the channel estimation techniques have been used based on mathematical models where error between the exact and predicted channel impulse response is minimized. Such computations are made on the noisy version of the received signal at the receiver. Therefore, the input signal and the corresponding received version are available, and without the information about the noise, the channel state is determined. The performance of such estimation schemes can be further improved when deep learning models are used. If the deep neural network is provided with a large dataset of input and output signals, it can be supervised to estimate the channel impulse response with higher accuracy. In fact, CNN can be trained by using time and frequency domain characteristics of the channel for predicting the response. Some of the contributions in the radio communication for channel estimation can be found in [35–39]. However, very less work has been presented for estimation of FSO signals. One such attempt for channel estimation in FSO systems has been studied in [40]. It is upon researchers to utilize more from deep learning models in representation of complex nature of optical signal fluctuations.

8.5.9 Network Traffic Prediction

Network traffic prediction in FSO communication systems is important for ensuring efficient resource management and optimizing network performance. The inherent challenges of FSO systems, such as their susceptibility to atmospheric turbulence and alignment requirements, necessitate sophisticated predictive mechanisms to maintain robust communication links. Deep learning, with its advanced data processing capabilities, offers a powerful approach to address these challenges.

Deep learning models, particularly those employing LSTM networks and CNNs, excel in capturing temporal and spatial dependencies within complex datasets. In the context of optical wireless systems, these models can be trained on recorded traffic data, including metrics such as error rates, data rate fluctuations, and atmospheric parameters. By learning from these patterns, deep learning models can predict future network traffic with high accuracy, enabling proactive management of network resources.

The predictive insights gained from deep learning models facilitate dynamic adjustment of transmission parameters, such as modulation schemes and power levels, to preemptively mitigate potential disruptions caused by traffic surges or adverse weather conditions. This proactive approach enhances the reliability and efficiency of FSO systems, ensuring sustained high-quality communication links. Furthermore, deep learning-based traffic prediction supports intelligent load balancing across the network, preventing congestion and optimizing the utilization of available bandwidth. This capability is particularly critical in scenarios where FSO systems are integrated with terrestrial networks, as it ensures seamless data flow and enhances overall network resilience.

8.5.10 Fault-Detection and Self-Healing Networks

A foundation for more robust, adaptive, and resilient optical communication networks can be established using deep learning models. Fault detection and self-healing mechanisms are critical components in the regulation and maintenance of FSO systems. Although, FSO systems are known for their sensitivity to environmental variations and physical obstructions, for future wireless communication systems, the ability of systems to detect and rectify faults on their own becomes a pivotal feature. Using vast datasets, the deep learning models excel at identifying patterns and anomalies. These models can be trained on a multitude of operational data, including signal strength, bit error rates, and environmental conditions. In real time, the deep learning models are more than equipped to detect faults in system such as signal degradation or alignment issues. The identification of subtle anomalies that traditional methods might overlook can be easily detected by deep learning algorithms due to their capability to handle high-dimensional data. Upon detection of a fault, deep learning-driven self-healing mechanisms can be activated. These mechanisms involve dynamic reconfiguration of the network to maintain service continuity. FSO systems can improve their fault prediction and recovery strategies over time, by continuously learning from operational data, leading to increasingly reliable communication systems. This capability is crucial for applications requiring high availability and low latency, such as data centers, 5G backhaul, and critical communication networks.

8.5.11 Localization and Positioning

Deep learning techniques offer innovative solutions to enhance the accuracy and reliability of critical tasks with their advanced pattern recognition and predictive capabilities. The accurate localization and positioning are challenged by factors such as atmospheric turbulence, physical obstructions, and the mobility of nodes, especially in dynamic environments in FSO systems. For proper and effective functionality of optical wireless systems, it is very important to utilize deep learning models for accurately positioning the various devices. Conventional methods have limited precision and susceptibility to environmental interferences and therefore cannot be utilized in the localization and positioning of FSO nodes. Such challenges can be met by various deep learning models such as RNN and CNN by processing and analyzing extensive datasets comprising signal strength indicators, received signal patterns, and environmental variables. Importantly, deep learning models can be trained to predict the precise location and alignment of FSO transceivers by learning from historical data and real-time inputs. Deep learning models can develop a comprehensive understanding of the spatial environment and therefore dynamically alter the positioning of FSO transceivers to maintain optimal alignment. This is a very important aspect of FSO systems due to the fact that these systems are LoS form of communication. A significant decrease in the need for manual intervention and minimizing alignment errors can be achieved using integration of deep learning models into localization and positioning systems that enable real-time and automated adjustments. Alignment with high precision can be achieved in systems with UAVs and satellite for enhancing the reliability of the optical wireless systems. By continuously monitoring and analyzing positioning data, these models can identify patterns that precede alignment failures or signal degradation, allowing for proactive adjustments and maintenance. These improvements are crucial for the deployment and operation of robust and efficient FSO networks, capable of supporting the increasing demands of modern communication systems.

8.6 Conclusion

This chapter focused on conducting research literature review and came up with ideas that can be implemented in FSO systems using deep learning. The future wireless communication systems will serve the high bandwidth required systems, and therefore, it is important to find substitute for conventional RF systems which are bandwidth inefficient. The FSO systems are of utmost importance in this regard. Furthermore, the autonomous nature of the future wireless systems needs attention of the deep learning and machine learning models for solving basic engineering problems. The chapter discussed in detail about some of the challenges that need to be solved with the aid of deep learning models. The ideas presented in the chapter will help the research community to find solutions to the practical problems that will be faced in the FSO communication networks. The various problems can be handled with different kinds of machine learning models based on the complexity of the problem.

References

1. C. Zheng, S. Yu, W. Gu, A SVM-based processor for free-space optical communication, in *2015 IEEE 5th International Conference on Electronics Information and Emergency Communication* (2015), pp. 30–33
2. K.S. Babatunde, F.A. Ibikunle, M.O. Arowolo, A.J. Alabi, E.A. Jiya, O.P. Kehinde, Machine learning model for classifying free space optics channel impairments, in *2022 5th Information Technology for Education and Development (ITED)* (2022)
3. J. Thrane, J. Wass, M. Piels, J.C.M. Diniz, R. Jones, D. Zibar, Machine learning techniques for optical performance monitoring from directly detected PDM-QAM signals. J. Lightwave Technol. **35**(4), 868–875 (2017)
4. A.S. Tan, H. Wymeersch, P. Johannisson, E. Agrell, P. Andrekson, M. Karlsson, An ML-based detector for optical communication in the presence of nonlinear phase noise, in *2011 IEEE International Conference on Communications (ICC)* (2011), pp. 1–5
5. M.P. Bart, N.J. Savino, P. Regmi, L. Cohen, H. Safavi, H.C. Shaw, S. Lohani, T.A. Searles, B.T. Kirby, H. Lee, R.T. Glasser, Deep learning for enhanced free-space optical communications. Mach. Learn.: Sci. Technol. **4**, 045046 (2023)
6. S. Song, Y. Liu, J. Wu, T. Wu, L. Zhao, L. Guo, Demonstration of intelligent hybrid FSO/RF system based on enhanced GRU prediction and real-world meteorological dataset. J. Lightwave Technol. **40**, 7048–7059 (2022)
7. W. Chen, M. Zhang, D. Wang, Y. Zhan, S. Cai, H. Yang, Z. Zhang, X. Chen, D. Wang, Deep learning-based channel modeling for free space optical communications. J. Lightwave Technol. **41**(1), 183–198 (2023)
8. B. Karanov, M. Chagnon, F. Thouin, T.A. Eriksson, H. Bülow, D. Lavery, P. Bayvel, L. Schmalen, End-to-end deep learning of optical fiber communications. J. Lightwave Technol. **36**(20), 4843–4855 (2018)

9. Z.-R. Zhu, J. Zhang, R.-H. Chen, H.-Y. Yu, Autoencoder-based transceiver design for OWC systems in log-normal fading channel. IEEE Photonics J. **11**(5), 1–12 (2019)
10. X. Liu, Z. Wei, A. Pepe, Z. Wang, H.Y. Fu, Autoencoder for optical wireless communication system in atmospheric turbulence, in *2020 Opto-Electronics and Communications Conference (OECC)* (2020), pp. 1–3
11. D. Wang, Y. Song, J. Li, J. Qin, T. Yang, M. Zhang, X. Chen, A.C. Boucouvalas, Data-driven optical fiber channel modeling: A deep learning approach. J. Lightwave Technol. **38**(17), 4730–4743 (2020)
12. H. Yang, Z. Niu, S. Xiao, J. Fang, Z. Liu, D. Fainsin, L. Yi, Fast and accurate optical fiber channel modeling using generative adversarial network. J. Lightwave Technol. **39**(5), 1322–1333 (2021)
13. F. Meng, P. Chen, L. Wu, X. Wang, Automatic modulation classification: A deep learning enabled approach. IEEE Trans. Veh. Technol. **67**(11), 10760–10772 (2018)
14. L. Huang, W. Pan, Y. Zhang, L. Qian, N. Gao, Y. Wu, Data augmentation for deep learning-based radio modulation classification. IEEE Access **8**, 1498–1506 (2020)
15. Y. Wang, J. Wang, W. Zhang, J. Yang, G. Gui, Deep learning-based cooperative automatic modulation classification method for MIMO systems. IEEE Trans. Veh. Technol. **69**(4), 4575–4579 (2020)
16. S.-H. Kim, C.-B. Moon, J.-W. Kim, D.-S. Kim, A hybrid deep learning model for automatic modulation classification. IEEE Wireless Commun. Lett. **11**(2), 313–317 (2022)
17. Y. Ni, S. Peng, L. Zhou, X. Yang, Blind identification of LDPC code based on deep learning, in *2019 6th International Conference on Dependable Systems and Their Applications (DSA)* (2020), pp. 460–464
18. P. Henarejos, M.Á. Vázquez, Decoding 5G-NR communications VIA deep learning, in *ICASSP 2020—2020 IEEE International Conference on Acoustics, Speech and Signal Processing (ICASSP)* (2020), pp. 3782–3786
19. Y. Wang, Z. Zhang, S. Zhang, S. Cao, S. Xu, A unified deep learning based polar-LDPC decoder for 5G communication systems, in *2018 10th International Conference on Wireless Communications and Signal Processing (WCSP)* (2018), pp. 1–6
20. Y. Wang, S. Zhang, C. Zhang, X. Chen, S. Xu, A low-complexity belief propagation based decoding scheme for polar codes—decodability detection and early stopping prediction. IEEE Access **7**, 159808–159820 (2019)
21. Y.E. Sagduyu, Y. Shi, T. Erpek, IoT network security from the perspective of adversarial deep learning, in *2019 16th Annual IEEE International Conference on Sensing, Communication, and Networking (SECON)* (2019), pp. 1–9
22. Md.A. Rahman, M.S. Hossain, A deep learning assisted software defined security architecture for 6G wireless networks: IIoT perspective. IEEE Wirel. Commun. **29**(2), 52–59 (2022)
23. L. Yang, J. Li, L. Yin, Z. Sun, Y. Zhao, Z. Li, Real-time intrusion detection in wireless network: A deep learning-based intelligent mechanism. IEEE Access **8**, 170128–170139 (2020)
24. Y. Li, Y. Zuo, H. Song, Z. Lv, Deep learning in security of Internet of Things. IEEE Internet Things J. **9**(22), 22133–22146 (2022)
25. M.I. Aziz Zahed, I. Ahmad, D. Habibi, Q.V. Phung, Content caching in industrial IoT: Security and energy considerations. IEEE Internet Things J. **7**(1), 491–504 (2020)
26. L. Liang, H. Ye, G. Yu, G.Y. Li, Deep-learning-based wireless resource allocation with application to vehicular networks. Proc. IEEE **108**(2), 341–356 (2020)
27. K.I. Ahmed, H. Tabassum, E. Hossain, Deep learning for radio resource allocation in multi-cell networks. IEEE Netw. **33**(6), 188–195 (2019)
28. W. Lee, R. Schober, Deep learning-based resource allocation for device-to-device communication. IEEE Trans. Wirel. Commun. **21**(7), 5235–5250 (2022)
29. R. Dong, C. She, W. Hardjawana, Y. Li, B. Vucetic, Deep learning for radio resource allocation with diverse quality-of-service requirements in 5G. IEEE Trans. Wirel. Commun. **20**(4), 2309–2324 (2021)
30. P. Yu, F. Zhou, X. Zhang, X. Qiu, M. Kadoch, M. Cheriet, Deep learning-based resource allocation for 5G broadband TV service. IEEE Trans. Broadcast. **66**(4), 800–813 (2020)

31. J. Guo, C. Yang, Predictive resource allocation with deep learning, in *2018 IEEE 88th Vehicular Technology Conference (VTC-Fall)* (2018), pp. 1–7
32. A.K. Nandi, E.E. Azzouz, Algorithms for automatic modulation recognition of communication signals. IEEE Trans. Commun. **46**(4), 431–436 (1998)
33. A.K. Nandi, E.E. Azzouz, Automatic analogue modulation recognition. Signal Process. **46**(2), 211–222 (1995)
34. M. Dahal, M. Vaezi, Deep reinforcement learning for interference management in millimeter-wave networks, in *2022 56th Asilomar Conference on Signals, Systems, and Computers* (2022), pp. 1064–1069
35. M. Soltani, V. Pourahmadi, A. Mirzaei, H. Sheikhzadeh, Deep learning-based channel estimation. IEEE Commun. Lett. **23**(4), 652–655 (2019)
36. H. Ye, G.Y. Li, B.-H. Juang, Power of deep learning for channel estimation and signal detection in OFDM systems. IEEE Wireless Commun. Lett. **7**(1), 114–117 (2018)
37. Q. Hu, F. Gao, H. Zhang, S. Jin, G.Y. Li, Deep learning for channel estimation: Interpretation, performance, and comparison. IEEE Trans. Wirel. Commun. **20**(4), 2398–2412 (2021)
38. H. Huang, J. Yang, H. Huang, Y. Song, G. Gui, Deep learning for super-resolution channel estimation and DOA estimation based massive MIMO system. IEEE Trans. Veh. Technol. **67**(9), 8549–8560 (2018)
39. Q. Bai, J. Wang, Y. Zhang, J. Song, Deep learning-based channel estimation algorithm over time selective fading channels. IEEE Trans. Cognit. Commun. Networking **6**(1), 125–134 (2020)
40. A. Upadhya, GAN based channel estimation and interference cancellation for mixed RF/FSO cooperative relaying systems. Phys. Commun. **61**, 102199 (2023)

Chapter 9
Deep Learning Enabled Channel Estimation for FSO Systems

9.1 Introduction

The remarkable expansion of wireless communication systems has given rise to the development of all five generations, spanning from 1G (first generation) to 5G (fifth generation). Yet, according to established timelines, the ongoing focus is on the research and development of the sixth generation (6G) of wireless and mobile communication systems. The primary goal is to cater to the substantial demands for data traffic. These 6G systems aim to leverage exceptionally large bandwidth, particularly in the terahertz (THz) band, and integrate artificial intelligence (AI). Evidently, the concept of "connected intelligence" serves as the foundational principle for the upcoming era of wireless communication systems.

Meeting the demands for massive connectivity, particularly in the high unlicensed bandwidth, poses a challenge for conventional RF systems and spectrum alone. To address this, the mmWave range has been considered as an alternative. However, reliance on mmWave communication may introduce additional regulatory constraints, potentially falling short of fulfilling the intended purpose. Consequently, the optical band of communication, particularly FSO communication systems, has emerged as a significant contender. Operating in the infrared wavelength of the spectrum, FSO communication systems enable operators to deliver bandwidth capable of supporting trillions of connected wireless devices.

FSO systems possess properties that make them well suited for meeting the demanding requirements of secure, ultrareliable, low-latency communication in 6G systems. Additionally, addressing the needs of self-configuration, context awareness, and cellular aggregation in 6G systems can be accomplished through the integration of AI, particularly leveraging deep learning models. The inherent capacity of deep learning models to comprehend datasets proves valuable in facilitating device-to-device communication and configuring heterogeneous networks within the 6G framework. The combination of deep learning and FSO

A. Upadhya et al., *Free-Space Optical Communication Systems for Next Generation Networks*, https://doi.org/10.1007/978-3-031-74805-9_9

systems demonstrates the capability to fulfill the stringent requirements of the next generation of wireless and mobile communication systems.

The present chapter focuses on implementing channel estimation method for FSO systems. The point of discussion in this chapter is in accordance with the various problems that are inherent in the modeling of the optical wireless channel. The research attempts to use the high-dimension feature of the deep learning models to represent the complex varying FSO channel.

9.2 State-of-the-Art Deep Learning Models

In recent years, the modeling of wireless communication channels at the physical layer has undergone advancements through various methodologies. It is expected in the 6G communications to utilize automated system architectures based on deep learning models to deploy radio access networks (RANs) for Internet of Things (IoT). For implementing femto-cell architectures, the utility of FSO networks will be immense due to the provision of large bandwidth on the optical links. An amalgamation of FSO and deep learning models can provide solutions to the AI-based high data rate RAN implementation.

In order to correctly model the channel, the main issue of the representation of complex varying wireless optical channels using mathematical models needs to be resolved. A better way of representing the channel estimates can be the utilization of the deep learning models. Once trained with real-time data, the model will be able to comprehend the network. A novel approach involving deep learning shows promise in offering an improved means of conceptualizing the challenges associated with channel modeling [1]. The intricate nature of wireless channel modeling encompasses diverse signal characteristics influenced by phenomena such as diffraction, scattering, reflection, attenuation, misalignment between the transmitter and receiver, and atmospheric factors. This complexity poses a significant challenge in accurately representing communication channels. While it is almost imperative to account for all relevant signal features in channel modeling, the inherent intricacy often leads researchers to employ various approximations, neglecting some key aspects of wireless signal behavior during analysis. Prioritizing the inclusion of all factors affecting signal quality is crucial. The forthcoming era of mobile communication demands a reevaluation of conventional channel modeling approaches, with a focus on leveraging the learning capabilities inherent in deep learning models.

In recent years, supervised machine learning methods (e.g., K-nearest neighbors, Bayes' theorem, decision tree, support vector machine, random forest, neural network, convolutional neural network, recurrent neural network, long short-term memory network, and gated recurrent unit network), unsupervised machine learning methods (e.g., association rules, K-means, density-based spatial clustering of applications with noise, hierarchical clustering, deep belief networks, deep Boltzmann machine, autoencoder, denoising autoencoder, etc.), reinforcement learning methods (e.g., generative adversarial network, deep Q network, trust region policy optimization, etc.), and federated learning methods have been applied to security and communication networks. Additionally, time series approaches such as the local outlier factor, cumulative sum, and adaptive online thresholding have been suggested to capture time series characteristics of abnormal behaviors to prevent cyberattacks and system malfunctions.

The importance of Shannon's work relates to the capacity offered by the wireless channel based on various efficient coding techniques. Even if a feedback is taken from the receiver to the transmitter, the main challenge lies in the utilization of linear coding schemes which are generally limited to suboptimal performance in terms of capacity. Therefore, a more practical yet robust way of channel coding (and decoding) that can use nonlinearity in the feedback has been advocated in the past using deep learning models. Historically, the main aim of channel coding schemes is to add redundancy to the information which the receiver can utilize to harness the correct information even in the presence of noise. Linear block codes and convolutional coding schemes have been combined to present turbo codes, which were used during 3G and 4G systems. For 5G systems, the complicated requirements of low latency and high bandwidth have advocated the use of low-density parity codes (LDPCs). The decoding delay can be minimized once the deep learning model is used to decode the polar coded schemes. Along with small delays, the property of automated performance near the enhanced mobile broadband (eMBB) is one of the key benefits of the deep learning models for decoding. Very recently, the deep learning models have been used to perform channel coding for wireless systems.

Deep learning models have been utilized for solving the complex problem of channel coding in the past. For instance, the work in [2] presents the first family of codes obtained via deep learning, which outperforms state-of-the-art codes designed over several decades of research. By integrating information theoretic insights with RNN-based encoders and decoders, the novel codes achieve three orders of magnitude improvement in reliability and a 3 dB gain in terms of SNR. Deep learning for channel decoding has been studied in work [3]. On the other hand, the researchers in [4] have implemented RNN-based approach for decoding the linear block codes. More related studies in this arena have been explored and elaborated in reference [5].

The mature technology of deep learning models, known for generating automatic solutions in image processing, biomedical science, and language processing (NLP), is witnessing continuous innovation with the introduction of various new algorithms. Recently, these models have extended their application to developing solutions for wireless communication technologies. However, their utilization for modeling the physical layer in wireless communication systems has been limited thus far. Notably, studies contributing to the use of deep learning models in wireless communication include references such as [1, 3, 4, 6–12]. In these works, the primary objective has been to leverage the universal approximation feature of deep learning algorithms to adapt to the complex and variable nature of wireless communication channels

[13]. It is worth noting that these studies have primarily focused on conventional RF systems.

9.3 Need for Deep Learning in FSO System

The necessity to deploy communication systems in novel frequency bands has resulted in the advancement of FSO systems [14]. In these systems, the transmission of information from the transmitter to the receiver occurs wirelessly over the optical link. However, this mode of information transfer introduces new challenges that must be addressed. Firstly, the communication link must establish a LoS between the telescope and the aperture. Any slight deviation in the orientation of the optical beam toward the receiver aperture results in what is termed as pointing error [15]. This misalignment fading can lead to significant information loss in FSO links. Additionally, optical wireless links are inevitably affected by free space path loss. Nevertheless, the most influential factor contributing to signal degradation is atmospheric turbulence—a unique characteristic of FSO links caused by fluctuations in refractive indices within the medium. To capture the impact of turbulence-induced fading, various mathematical models have been employed in the literature [16–20]. However, these models pose limitations in analysis due to mathematical intricacies. Despite this, analyses employing these models often rely on various assumptions, introducing deviations from practical observations in the results.

Detecting the characteristics of signal fading allows for mitigation of various channel variations occurring in FSO inks. Additionally, deploying a real-time beam tracking system and adjusting the telescope/aperture orientation can minimize the impact of beam wander, including pointing error. When real-time CSI from the FSO link is gathered and fed into deep neural networks (DNNs), it becomes possible to predict future channel conditions. This prediction can be instrumental in mitigating the effects of turbulence and pointing error. The very recently proposed generative adversarial networks (GANs) and long short-term memory (LSTM) models can be used for channel estimation in FSO systems. In this chapter, the DNN model has been used for channel estimation on the FSO link.

9.4 Artificial Neural Networks

The ANN architectures are decentralized networks of computing elements designed to emulate biological neural systems. They serve the purpose of identifying relationships within input data and predicting connections within newly introduced datasets. These algorithms are employed for nonlinear statistical modeling, offering an alternative to logistic regression in forecasting medical outcomes. Essentially, they function as computer simulations of biological parallel distributed processing systems, capable of executing intricate pattern recognition tasks and constructing models that mimic brain function. ANNs may have a single layer (perceptron) or multiple layers (multilayer perceptron). The connections between neurons have weights and biases that are adjusted during the learning process. Generally, the gradient descent algorithm is implemented to compare the output of the model with the input, and the weights and biases are adjusted accordingly to minimize the error in each iteration. The sequence of altering the weights and biases to decrease the error between the output and the input is called one epoch. Training an ANN involves adjusting the weights based on the input data and desired output using algorithms like backpropagation. ANNs can learn complex relationships in data. The number of epochs required for training the model depends on the number of samples and the complexity of the underlying problem. The inputs are called the features, while the outputs are called the labels. Generally, the ANN model consists of three layers. The first layer is the input layer, and the number of neuron units in this layer is equal to the number of features that are present to train the model. The output layer is the third layer which consists of the number of neurons equal to the number of labels. The layer in between the input and output layers is said to be the hidden layer. The number of neurons in the hidden layer and the number of hidden layers are problem-dependent and vary based on the complexity of the model with which the ANN model is being trained. Such ANN models with features and labels are called to be supervised models. During the training process, the neurons are connected through synaptic weights and biases, and the output of neuron is passed through the activation function. The role of the activation function is to activate the neuron and is generally a mathematical function. Rectified Linear Unit (ReLU), sigmoid, linear, and leaky ReLU are some of the most preferred activation functions.

However, in some cases, it is desirable to extract the important information from the input data only and utilize these features for classification. This type of ANN comes under the unsupervised models. There is a third category of ANN models called the reinforcement learning models which have the capability to learn to solve the underlying problems based on the rewards that each action gets in an environment to accomplish the desired outcome. For this, it is necessary to provide the environmental conditions to the model and let the model learn from its mistakes and get the job done. The training of reinforcement learning algorithms is based either on the Q function, where the model can only access a definite number of states, or on maximizing a policy function, where Stochasticity is used to determine the optimal actions, providing a hypothesis based on an infinite number of possible states. The key components and concepts associated with reinforcement learning models are:

- Agent: the entity or system that takes actions within the environment
- Environment: the external system or context in which the agent operates and receives feedback
- State: a representation of the current situation or configuration of the environment
- Action: the decisions or moves made by the agent that affect the state of the environment

- Reward: numeric feedback is provided to the agent after each action, indicating the desirability of the action
- Policy: the strategy or a set of rules that the agent uses to determine its actions based on the current state
- Value Function: a function that estimates the expected cumulative reward from a given state or state-action pair
- Exploration and Exploitation: balancing the trade-off between exploring new actions and exploiting known actions to maximize rewards

  Discount Factor: a parameter that discounts future rewards, emphasizing the importance of immediate rewards in decision-making

9.4.1 Comparison Between ANN and DNN Models

ANNs and deep learning models closely relate, yet they differ in certain aspects:
Scope and Depth:

ANNs: ANNs are a specific computational model inspired by the human brain's neural networks. They consist of interconnected nodes, or neurons, in layers, receiving, processing, and outputting signals.

Deep Learning: a subfield of machine learning; deep learning focuses on brain-inspired algorithms, notably using ANNs with numerous layers to learn data representations hierarchically.

Hierarchy:

ANNs: With one or multiple layers, traditional ANNs flow information from input through hidden layers to output.

Deep Learning: Specifically, deep learning utilizes ANNs with many hidden layers, often more than three, to learn complex features from input data.

Complexity and Representation Learning:

ANNs: Capable of learning complex data patterns, ANNs may struggle with large or high-dimensional datasets, requiring manual feature engineering.

Deep Learning: Excels at learning intricate patterns from vast data without hand-crafted features, automatically capturing complex relationships and variations in input.

Applications:

ANNs: Applied in image and speech recognition, time series forecasting, and pattern recognition.

Deep Learning: Highly successful in image and speech recognition, natural language processing, autonomous driving, drug discovery, and more. Its prowess with extensive data and complex patterns has cemented its role in modern AI research and applications.

The deep learning models are more complex and generic when compared to the ANN models. The main difference comes in the number of hidden layers. Generally, deep learning models are used to model more complex input-output relationships where it is not possible to work with a less number of hidden units. The large number of hidden layers equips the deep learning models to learn the hierarchical representations of data. The manual feature extraction can be eliminated by the implementation of deep learning models. Domains like computer vision, natural language processing, wireless communication analysis, and image processing among others have been implemented using deep learning models.

9.4.2 Conventional Methods Versus Deep Learning for Channel Estimation

Channel estimation using conventional methods can be time-consuming and labor-intensive, requiring domain experts to design features that represent communication channel characteristics. This approach may not fully capture channel complexity. Conventional methods may struggle to adapt to dynamic or nonstationary channel conditions due to being based on specific assumptions about channel models and environmental factors. In complex or highly dynamic environments, performance degradation may occur, impacting reliability and data throughput in wireless communication systems. Additionally, sensitivity to noise and interference might affect the accuracy of channel estimation in traditional methods, leading to estimation errors.

On the other hand, deep learning models provide various benefits. These models can automatically learn hierarchical data representations from raw input, eliminating the need for manual feature engineering. They can adaptively capture complex relationships and variations in the communication channel without predefined features. Deep learning models are flexible and can adapt to diverse channel conditions and environments. By learning from large-scale datasets, they continuously improve performance and adjust to changing channel dynamics. These models can effectively filter out noise and interference in the communication channel, enhancing robustness and reliability in challenging environments. Deep learning models have shown scalability and generalization across different communication scenarios and environments. Once trained, they can be deployed across various wireless communication systems with minimal adaptation, making them a versatile solution for channel estimation tasks.

9.5 FSO Channel Estimation Using Deep Learning

In this section, channel estimation on the FSO link has been presented using the deep learning model. Suppose a laser diode is used to transmit optical signal through wireless media with signal undergoing atmospheric turbulence and pointing error. The FSO system transmits the information symbol x over an optical wireless channel with turbulence represented as I, which is also perturbed due to AWGN represented by n. The received signal at the receiver aperture after being transformed by photodetector can be written as

$$y = xI + n \tag{9.1}$$

Now, due to noisy channel and the effect of atmospheric turbulence and pointing error on the FSO link, the symbol x is impacted and y is a noisy version of the same. Various studies in the past have worked on channel estimation methods which generally involve the statistical characterization of the fading distribution over which the information is transmitted. However, in recent years, the deep learning models have been implemented to solve this problem. If trained in an efficient manner, the deep learning model can predict the channel estimate I for corresponding values of y, avoiding complex statistical mathematical derivations for estimation.

The deep learning model in the present work has been trained online with raw data of input vectors and received output signal. The model does not have any information about the channel and the noise component. Only using the input signal with received output, the channel estimator based on deep learning model has the capability to be trained in supervision to estimate the channel response of the optical wireless channel. The offline training of the model has been done using the Keras library. The first operation of the training processing is the preprocessing of the input (features) and output (labels). This has been done using the *MinMaxScaler* function from *sklearn* library [21]. The deep learning model consists of three layers. The first layer is the input layer. In the present case, there are four neurons in the input layer. The activation of neurons in the input layer is done using the ReLU function. For better training of the model, the output of the first layer is also normalized using *BatchNormalization*. Another layer termed the hidden layer with four neurons and ReLU activation is used, while the output layer is activated using the sigmoid function with one neuron for providing the channel estimate on the FSO link. The backpropagation algorithm is used to update the weights and biases of the deep learning model with 50 epochs. The performance of the trained model and channel estimation is depicted in Fig. 9.1.

The main purpose of the present study is to carry out channel estimation for FSO link using deep learning model. For demonstration, deep learning model (DNN) is used as shown in Fig. 9.2. The dataset has been generated using MATLAB software. The model is trained with dataset of $X_i = x_i, y_i$ as features and $Y_i = I_i$ labels, where $i \in N_1$. These N_1 points are obtained from total N length of dataset by splitting 0.7 and 0.3 as training and testing data. The trained model is then used to test for the N_2 data points and validated with the actual labels available in offline mode. Once deep learning model is trained with high accuracy, it can be employed in online mode for the purpose of channel estimation in real time. This means that $N = N_1 + N_2$.

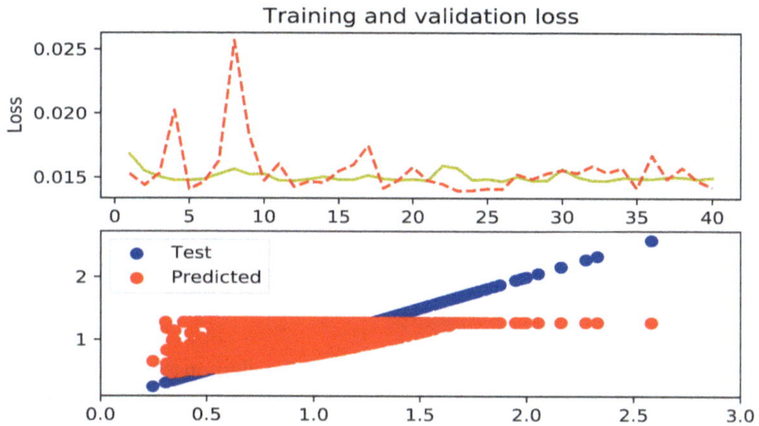

Fig. 9.1 The performance of the deep learning model in terms of validation loss and correlation between the actual channel impulse response and the estimated version

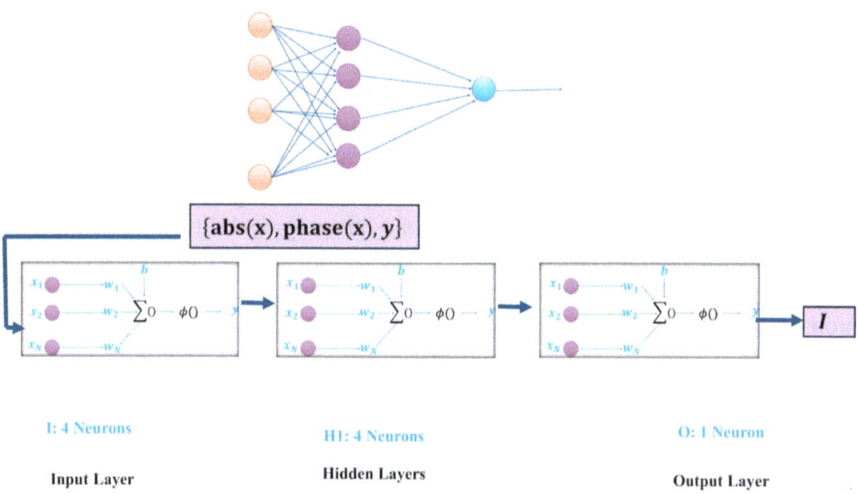

Fig. 9.2 The experiment setup for channel impulse response estimation

For the purpose of evaluation of the performance of the model, mean square error (MSE) and root mean square error (RMSE) are calculated. The MSE of the experiment is defined as

$$MSE = \frac{1}{N_2} \sum_{i=1}^{N_2} (I_{test} - I_{pred})^2 \tag{9.2}$$

where I_{test} and I_{pred} are samples of the channel response from the test and predicted values. On the other hand, the RMSE values can be evaluated using the relation

$$RMSE = \sqrt{\frac{1}{N_2}\sum_{i=1}^{N_2}(I_{test} - I_{pred})^2} \tag{9.3}$$

The corresponding dB values can be calculated as $MSE_{dB} = 10log(MSE)$ and $RMSE_{dB} = 10log(MSE)$.

9.6 Results

The results of trained deep learning model can be demonstrated not only using scattering plot but also on the basis of some performance metrics. The outage probability and ergodic capacity have been plotted using the test and predicted data in Figs. 9.3 and 9.4 for this purpose. The deep learning model has been trained with MSE of 0.0009043 which corresponds to −30.47 in dB. Another performance metric used for evaluation is the RMSE in dB which has been obtained as −5.54 dB. The outage probability has been demonstrated for two values of threshold SNR on the optical wireless link. The threshold SNR values have been opted as $\gamma_{th} = [-5, 5]$ dB. The analysis has been done based on the calculation of outage probability using the test dataset values of channel impulse response and the outage probability evaluated using the DNN predicted values of channel. As the threshold SNR increases, the system experiences higher outage probability. The trained model is able to comprehend the outage probability values which are obtained using the real-time test dataset. The trained model can be placed in the real world to predict the channel response and evaluate the corresponding outage probability values.

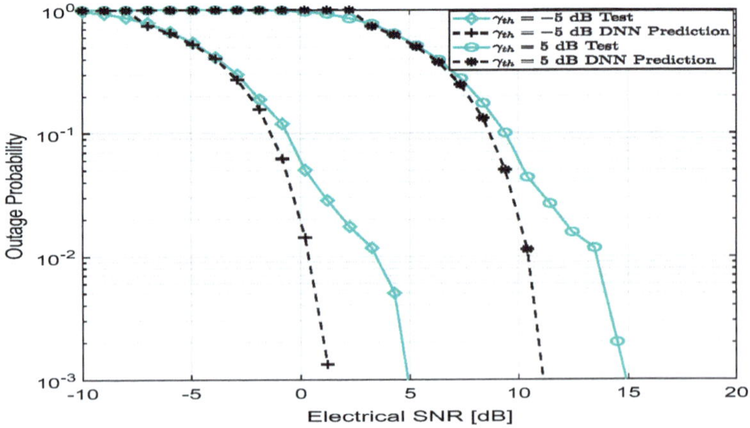

Fig. 9.3 Outage probability of an all FSO link for two different combinations of threshold SNR

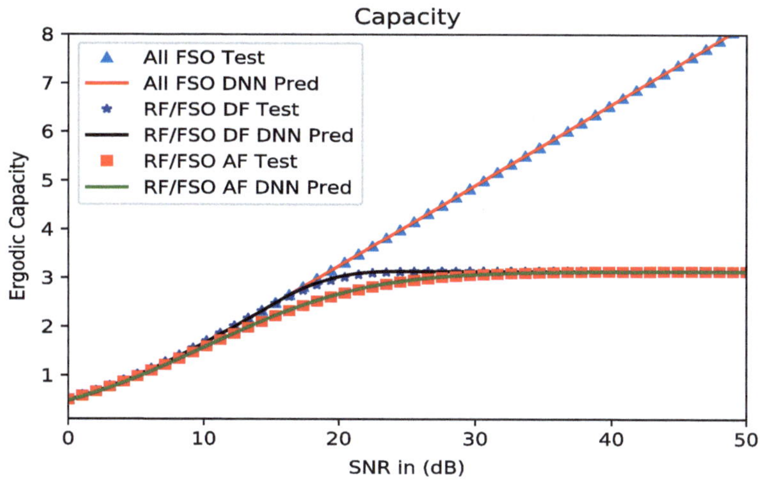

Fig. 9.4 Ergodic capacity of an FSO link compared with DF and AF counterparts for fixed RF SNR

On the other hand, ergodic capacity analysis is predicted using the trained deep learning model to compare the relaying strategy with single-hop transmission. The SNR on the RF link has been fixed which has been assumed to be $\bar{\gamma}_{RF} = 10\,\text{dB}$. The ergodic capacity has been plotted against electrical SNR values of $0-50\,\text{dB}$. As the signal strength on the FSO link increases, the capacity increases. Also, single-hop relaying becomes promising for the set of model parameters used in the case. The DF relaying is also compared with AF relaying. The DF relaying provides larger capacity for low SNR values compared with AF relaying systems. The most important illustration in the plot is the fact that the test data is completely followed by the deep learning model prediction. The trained model can be effectively used to prediction of throughput of FSO systems of various kinds.

9.7 Conclusion

The world is using to solve a number of problems using deep learning models. In wireless communication networks, the optimization of the physical layer is based on channel characteristics. If the channel impulse response is available, it is possible to improve the system performance with the aid of signal processing techniques. The examination of channel in FSO systems using deep learning models represents a significant advancement in optical communication. By employing deep learning, innovative methods can be discovered to accurately predict channel characteristics, enabling FSO systems to adjust and excel in various dynamic environments. This chapter has delved into the detailed mechanisms of deep learning structures, their capacity to grasp intricate patterns from information, and their utilization in FSO channel estimation. Looking forward, the collaboration between deep learning and FSO technology shows potential for enhancing the dependability, efficiency, and scalability of optical communication systems. This progress clears the path for a future where high-speed, reliable, and adaptable FSO networks become essential components of our global communication infrastructure.

References

1. T. O'Shea, J. Hoydis, An introduction to deep learning for the physical layer. IEEE Trans. Cognit. Commun. Networking **3**(4), 563–575 (2017)
2. H. Kim, Y. Jiang, S. Kannan, S. Oh, P. Viswanath, DeepCode: Feedback codes via deep learning. IEEE J. Sel. Areas Inf. Theory **1**(1), 194–206 (2020)
3. S. Cammerer, T. Gruber, J. Hoydis, S.T. Brink, Scaling deep learning-based decoding of polar codes via partitioning, in *GLOBECOM 2017—2017 IEEE Global Communications Conference* (2017), pp. 1–6
4. E. Nachmani, E. Marciano, D. Burshtein, Y. Be'ery, RNN decoding of linear block codes. ArXiv, abs/1702.07560 (2017)
5. A. Ly, Y.-D. Yao, A review of deep learning in 5G research: Channel coding, massive MIMO, multiple access, resource allocation, and network security. IEEE Open J. Commun. Soc. **2**, 396–408 (2021)
6. M. Ibukahla, J. Sombria, F. Castanie, N.J. Bershad, Neural networks for modeling nonlinear memoryless communication channels. IEEE Trans. Commun. **45**(7), 768–771 (1997)
7. J. Sjoberg, Q. Zhang, L. Ljung, A. Benveniste, B. Delyon, P.-Y. Glorennec, H. Hjalmarsson, A. Juditsky, Nonlinear black-box modeling in system identification: a unified overview. Automatica **31**(12), 1691–1724 (1995). Trends in System Identification
8. M. Soltani, V. Pourahmadi, A. Mirzaei, H. Sheikhzadeh, Deep learning-based channel estimation. IEEE Commun. Lett. **23**(4), 652–655 (2019)
9. H. Ye, G.Y. Li, B.-H. Juang, Power of deep learning for channel estimation and signal detection in OFDM systems. IEEE Wireless Commun. Lett. **7**(1), 114–117 (2018)
10. Q. Hu, F. Gao, H. Zhang, S. Jin, G.Y. Li, Deep learning for channel estimation: Interpretation, performance, and comparison. IEEE Trans. Wirel. Commun. **20**(4), 2398–2412 (2021)
11. H. Huang, J. Yang, H. Huang, Y. Song, G. Gui, Deep learning for super-resolution channel estimation and DOA estimation based massive MIMO system. IEEE Trans. Veh. Technol. **67**(9), 8549–8560 (2018)
12. Q. Bai, J. Wang, Y. Zhang, J. Song, Deep learning-based channel estimation algorithm over time selective fading channels. IEEE Trans. Cognit. Commun. Networking **6**(1), 125–134 (2020)
13. K. Hornik, M. Stinchcombe, H. White, Multilayer feedforward networks are universal approximators. Neural Netw. **2**(5), 359–366 (1989)
14. Z. Ghassemlooy, W. Popoola, S. Rajbhandari, Optical Wireless Communications: System and Channel Modeling with MATLAB (CRC Press, New York, 2012)
15. A.A. Farid, S. Hranilovic, Outage capacity optimization for free-space optical links with pointing errors. J. Lightwave Technol. **25**, 1702–1710 (2007)
16. J.H. Churnside, S.F. Clifford, Log-Normal Rician Probability Density Function of Optical Scintillations in the Turbulent Atmosphere. J. Opt. Soc. Am. **4**, 1923–1930 (1987)
17. L.C. Andrews, R.L. Phillips, *Laser Beam Propagation Through Random Media*, 2nd edn. (SPIE Press, Bellingham, WA, 2005)
18. M. Al-Habash, L.C. Andrews, R.L. Phillips, Mathematical model for the irradiance probability density function of a laser beam propagating through turbulent media. Opt. Eng. **40**(8), 1554–1562 (2001)

19. A.J. Navas, J.M.G. Balsells, J.F. Paris, A.P. Notario, A unifying statistical model for atmospheric optical scintillation, in *Numerical Simulations of Physical and Engineering Processes, Ch. 8* (2011)
20. M.A. Kashani, M. Uysal, M. Kavehrad, A novel statistical channel model for turbulence-induced fading in free-space optical systems. IEEE/OSA J. Lightwave Technol. **33**(11), 2303–2312 (2015)
21. M.T. Hagan, H.B. Demuth, M.H. Beale, O.D. Jes'us, *Neural Network Design*, 2nd edn. (Martin Haga, USA, 2014)

Chapter 10
CNN and LSTM Deep Learning Models for Channel Estimation of Two-Way Relaying in the Presence of Hardware Impairment

10.1 Introduction

In the previous chapter, the conventional ANN deep learning model has been implemented for channel estimation of FSO link. The CNN deep learning models find their application in image-based problems. However, recently, CNNs have gained attention in the realm of wireless communication due to their feature of feature reduction. On the other hand, long short-term memory (LSTM) deep learning model presents an interesting way to cater even more input data. This makes LSTM deep learning models suitable for channel estimation. This chapter dwells deeper into modeling of optical wireless channel using CNN and LSTM models.

10.2 Channel Estimation Using CNN

Channel estimation plays an important role in wireless communication systems with an aim to precisely determine the characteristics of the communication channel between the transmitter and the receiver. The precision of channel estimators surprisingly influences the performance of multiple communication functionalities such as beamforming and interference reduction. Common methods for channel estimates typically use pilot signs and models to estimate channel parameters. However, these methods could face issues in handling the dynamic and nonlinear natures of wireless channels, especially in situations described by high mobility, packed multipath propagation, and selective-fading of frequency.

Deep learning CNN model has created a massive change in wireless communication systems lately. The last few years have seen a revolution in channel estimation process. CNNs have shown fantastic skills in understanding the complex connections between signals and channel settings, outpacing the boundaries of old-fashioned estimation methods. Leveraging spatial and timely correlations in

A. Upadhya et al., *Free-Space Optical Communication Systems for Next Generation Networks*, https://doi.org/10.1007/978-3-031-74805-9_10

wireless channels, channel estimation techniques based on CNN models offer improved accuracy, strength, and adaptability in different communication setups.

An essential utilization of CNN-based channel estimation lingers within millimeter-wave (mmWave) massive multiple-input multiple-output (MIMO) platforms. In the research work [1], a sole CNN structure fabricated for channel estimation in mmWave MIMO systems has been presented. This research stresses the efficacy of CNNs in seizing the spatial connections innate in mmWave systems, yielding in superior estimation execution relative to usual methods. Throughout the utilization of profound estimation methodologies, the proposed tactics permit precise and efficient channel estimation, establishing the scenario for the inclusion of mmWave MIMO platforms in subsequent radio networks.

Incorporating CNN-based channel estimation techniques has shown effective results in MIMO-OFDM systems aside from mmWave technology, commonly used in wireless communication standards such as LTE and LTE-advanced. The authors in [2] showcase a double CNN framework specifically designed for channel estimation in MIMO-OFDM setups. This framework capitalizes on the organizational features of OFDM signals and the diverse spatial aspects of MIMO systems to enhance estimation precision and robustness. By comprehensively estimating both the frequency and spatial field features of the channel, the suggested approach attains significant performance improvements compared to conventional techniques.

The exhumation of CNN-oriented channel estimation techniques among vehicular correspondences has faced hurdles owing to high-motion and varied spread circumstances, resulting in the requirement of new methods. In [3], researchers suggest an approach that exploits CNN for bettered channel estimation accuracy in vehicular communication frameworks via interpolation plan. This merging of CNNs in the interpolation procedure permits adaptability to varying channel situations, thereby enriching estimation operation and durability in active environments. Furthermore, [4] delves into frequency aspect in mmWave MIMO structures deploying CNN structure.

Moreover, the CNN realm-based channel assessment surpasses traditional communication scenes to blossoming territories like the IoT. The research in [5] reveals a groundbreaking OFDM encoder design incorporating CNN-dependent channel estimation for high-speed IoT utilization. Embedding the estimation of channels capabilities straight into the framework of communication systems, this method eases the placement and function of IoT devices on nautical ships while providing reliable and effective communication in demanding maritime environments.

10.3 LSTM for Channel Estimation

LSTM deep learning models have gained attention in wireless system because they catch time-linked relationships within the parameters in the channels. They can overcome the constraints of conventional methods of channel estimation. With fresh layouts and training, estimation techniques based on LSTM models provide superior accuracy, power, and adaptability, trustworthy communication backed up in numerous tough scenarios. Continuing research in this sector, the usage of LSTM-based estimation is poised to play a significant role in shaping forthcoming wireless communication systems.

The research paper presented in [6] has made a big contribution to LSTM-based channel estimation technique. It introduces ChanEstNet, a new deep learning system based on LSTM networks made for estimating channel in fast fading wireless systems. ChanEstNet outperforms regular methods in accuracy by taking advantage of channel time-dependent qualities, particularly in fast-changing surroundings led by high movement. The importance of this work is found in its ability to provide reliable channel estimation in vehicular environments, enabling strong communication in fast-speed atmospheres like vehicular communication and high-speed rail systems. LSTM-based channel estimations have been into numerous communication situations behind high-velocity scenarios, including high-frequency (HF) MIMO single carrier frequency domain equalization (SC-FDE) systems. The study in [7] presents a customized online LSTM-based channel estimation technique for HF MIMO SC-FDE systems. This strategy uses the enhanced memory capabilities of LSTM networks to dynamically predict channel parameters in real time, particularly in challenging HF propagation circumstances. Continuously updating channel estimates based on received data, this proposed method enhances the resilience and dependability of communication across HF channels that are susceptible to significant fading and interference.

Techniques relying on LSTM for estimating channels have been implemented in the context of vehicular communications employing mmWave technology. These communication systems encounter challenges due to high mobility and changing propagation conditions, which traditional estimation methods struggle to tackle. The research presented by the authors of [8] introduces a framework relying on deep learning for channel estimation and tracking in mmWave vehicular communications. This framework utilizes LSTM networks to understand the temporal changes in mmWave channels and monitor their variations over time. By integrating LSTM-based channel estimation into vehicular communication systems, this method boosts communication reliability and performance, enabling seamless connectivity for autonomous vehicles and smart transportation systems. By exploiting the temporal correlations in the channel, the proposed scheme accurately predicts future channel states, allowing proactive adaptation of transmission strategies to enhance communication security, and reliability. Furthermore, the paper [9] proposed an LSTM-based channel prediction scheme to mitigate the effects of imperfect CSI on secure massive MIMO communications. The LSTM-based channel estimation methods have been investigated in the context of secure massive MIMO communications, where accurate channel prediction is crucial for achieving secure and reliable communication under imperfect channel state information (CSI).

10.4 Convolutional Neural Networks

In the realm of deep learning, a CNN represents a kind of complicated neural networks mainly used for analyzing visual data. While traditional neural networks rely seriously on matrix multiplications, ConvNets differ by using a special method known as convolution. In mathematical terms, convolution is an operation applied to two functions, producing a third function that shows the transformation of one function through the other. In the human brain, visual processing happens quickly when encountering an image, with each neuron operating within its receptive field, intricately interconnected with others to cover the entire visual spectrum. Similarly, in CNN models, individual neurons process data only within their receptive fields, reflecting the biological vision system. These networks are organized to detect basic patterns initially (such as lines and curves) and advance to more intricate features like faces and objects. Through utilizing CNNs, the idea of vision can be expanded to computers. In solving the problems related to wireless communication systems, CNN deep learning models have been strongly favored by researchers in the past. The main reason for this is that the large number of sequential data can be converted to two-dimensional dataset and can be provided as input in the CNN model for trained and validated purposes. The convolutional operations are performed in the convolutional layers which capture spatial hierarchies and patterns in the input. The pertinent feature of the input data is the feature map which is used for understanding the relation among the inputs and outputs datasets. Typically consisting of three layers—convolutional, pooling, and fully connected—the convolutional layer is the fundamental component responsible for the network's computational workload. This layer conducts a dot product operation between two matrices: One matrix represents the learnable parameters, known as a kernel, and the other matrix represents a restricted portion of the receptive field. The kernel, spatially smaller than the image, extends deeper. Consequently, while the image comprises three (RGB) channels, the kernel's spatial dimensions are compact, yet its depth spans across all three channels.

To conduct the convolutional filtering, a pooling operation is needed for reduction of the dimensionalities done using down-sampling [10, 11]. Pooling techniques such as maximum pooling and average pooling is generally performed. Pooling aids in retaining essential information while diminishing computational complexities. The multidimensional datasets generated in the convolutional layers are made to be flattened through the fully connected flatten layers. On the other hand, the pooling layer substitutes the network's output at specific locations by computing summary statistics of nearby outputs. This aids in diminishing the spatial size of the representation, thereby reducing computational requirements and weights. The pooling operation is executed individually on each slice of the representation. Following are key components of CNN models:

Fully Connected Layer: Neurons within this layer exhibit full connectivity with all neurons in the preceding and succeeding layers, akin to regular Fully Connected Neural Networks (FCNNs). Hence, it can be computed conventionally via matrix multiplication followed by bias addition. *Nonlinearity Layers*: Given that convolution constitutes a linear operation, whereas images exhibit nonlinearity, nonlinearity layers are frequently positioned immediately after the convolutional layer to introduce nonlinearity to the activation map. Popular examples include Sigmoid, Tanh, and ReLU. *Flattening*: After several convolutional and pooling layers, the data undergoes flattening before traversing through fully connected layers. *Activation Functions*: ReLU is widely employed in CNNs to introduce nonlinearity by converting negative values to zero. Additionally, other activation functions like Sigmoid and Tanh are utilized in specific layers for particular tasks. *Convolutional Neural Network Architecture*:

Architectural Variants: Within the domain of CNNs, architectural diversity is evident through notable models such as LeNet-5, AlexNet, VGGNet, GoogLeNet, and ResNet. These models exhibit disparities in layer depth, inclusion of skip connections, and other design components. Transfer Learning: Leveraging pretrained CNN models, typically trained on comprehensive image datasets like ImageNet, facilitates their adaptation to specific tasks. This approach optimizes training time and resource utilization, enhancing efficiency in task-specific model development.

10.5 LSTM Deep Learning Models

Deep learning methods have been lately effectively applied in the sphere of wireless communication, with a specific focus on channel foretelling techniques employing deep learning algorithms. Most notably, LSTM is appearing as a proficient and superior version of recurring neural networks (RNNs). It adequately deals with the issue of long-term interdependencies faced in traditional RNNs and is widely useful in handling time series and vocal data [12].

LSTM, unlike RNNs, excels in capturing long-term dependencies within sequence data. They feature the ability to analyze and interpret sequence data kinds like temporal series, text data, and speech. Through using a memory cell beside gateway mechanisms, LSTMs control information flow, allowing selective retention or discarding of data as needed, hence minimizing issues like the vanishing gradient problem. Their adaptability has resulted in extensive adoption in diverse sectors, involving natural language processing, speech acknowledgment, and temporal series prediction.

The LSTM structure includes myriad LSTM units with an array of gates (input, output, and forget gates) that control the flow of facts into and out of the unit. The gates perform a vital function in particularly holding or discarding facts from preceding moments, permitting LSTM to efficiently seize lengthy-term dependencies within input statistics. In the realm of wireless communications, LSTM has been employed in distinct research initiatives, encompassing channel approximation, correlation scrutiny of wireless channel statuses, and diagramming of beamforming in mmWave backgrounds. Notably, there is research in the use of LSTM patterns to simulate FSO connections. Three types of gates subsist within an LSTM: the input gate, the forgetting gate, and the output gate. Every gate is actualized using sigmoid functions, bringing forth productions between 0 and 1. These gates are coached using backward propagation procedures throughout the network.

The input gate determines what information is stored in the memory cell, opening for important entries and closing for less relevant ones. The forgotten gate controls what information is thrown away from the memory cell, opening to throw out unnecessary details and closing to maintain necessary information. Moreover, the output gate determines what information impacts the LSTM's output, opening for significant data and closing for unimportant information. These gates flexibly change their openness relying on inputting and previous hidden states, enabling the LSTM to choose to keep or get rid of information, improving its capability to grasp long-term dependencies.

10.6 System Model

A two-way relaying network has been considered for demonstrating the channel estimation based on CNN and LSTM deep learning models as shown in Fig. 10.1. Two nodes communicate among themselves on RF and FSO links. Node N_1 sends information through relay node to node N_2. The FSO channel estimation is performed online using the CNN and LSTM models. The node N_1 sends symbol s_1 on RF link, while FSO link sends s_2 as carrier symbol. Both these signals

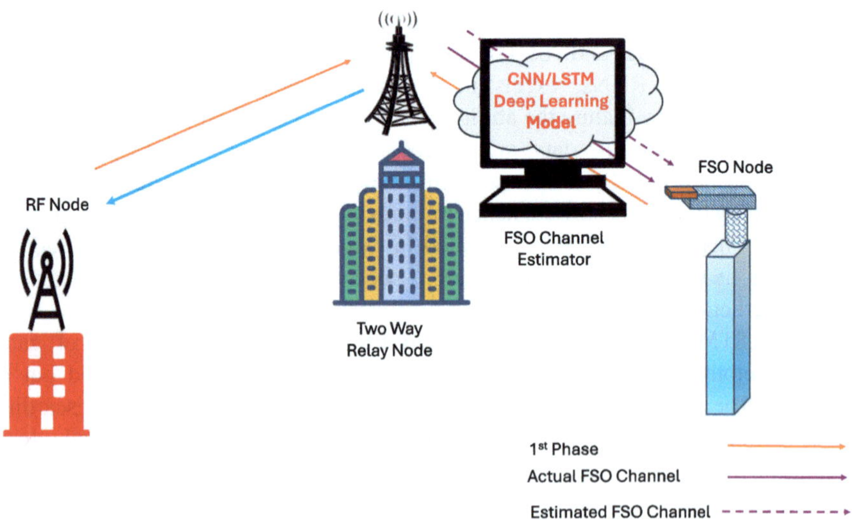

Fig. 10.1 System model for deep learning enabled channel estimation for FSO link in the presence of transmit and receive hardware impairment at relay node

are received by the relay node. However, the relay node operates with hardware impairment. The relay node receives signal which can be mathematically written as

$$y_r = h_r s_1 + (\eta I)^{r/2} s_2 + \mathcal{K}_r + n_1 \tag{10.1}$$

Here, h_r denotes the fading coefficient on the RF link modeled as Nakagami-m model, I is optical link turbulence, η is the optical-to-electrical converter, and $\mathcal{K}_r$ is the hardware impairment. The AWGN component is denoted as n_1. The aggregate hardware impairment is modeled as $\mathcal{K}_r \backsim \mathcal{N}(0, \kappa_r^2(|h_r|^2 P_1 + (\eta I)^r))$, where RF link transmits with power P_1. The variable gain applied by the relay node can be expressed as

$$G = \sqrt{\frac{P_r}{(|h_r|^2 P_1 + (\eta I)^r)(1 + \kappa_r^2) + \sigma_n^2}} \tag{10.2}$$

where P_r is the power by relay and AWGN variance is represented as σ_n^2.

The relay sends the information to both the links also with hardware impairment such that $\mathcal{K}_t \backsim \mathcal{N}(0, \kappa_t^2 P_r)$. Using the gain, the signal received at N_1 is expressed as

$$y_1 = (\eta I)^{r/2} h_r s_2 + G(\eta I)^r s_1 + G h_r (\mathcal{K}_r + n_1) + h_r \mathcal{K}_t + n_2 \tag{10.3}$$

Similarly, the received signal at the FSO link, transmitted by the RF link, can be devised as

$$y_2 = (\eta I)^{r/2} h_r s_1 + G h_r s_2 + G(\eta I)^{r/2} (\mathcal{K}_r + n_1) + (\eta I)^{r/2} \mathcal{K}_t + n_3 \tag{10.4}$$

It has been assumed that n_2 and n_3 are the AWGN factors on the N_1 to relay and N_2 to relay links. Assuming that self-interference can be canceled by the nodes, the signal-to-distortion-plus-noise ratio (SNDR) for each node can be expressed as follows:

$$SNDR_1 = \frac{\gamma_1 \gamma_2}{\frac{\gamma_1^2}{P_r} c + \gamma_1 \gamma_2 c + \gamma_2 P_1 b_1 + \gamma_1 (a_1 + P_1 b_1) + \frac{\sigma_n^4}{P_r}} \tag{10.5}$$

$$SNDR_2 = \frac{\gamma_1 \gamma_2}{\frac{\gamma_2^2}{P_r} c + \gamma_1 \gamma_2 c + \gamma_1 b_2 + \gamma_2 (a_2 + P_1 b_2) P_1 + \frac{\sigma_n^4}{P_1 P_r}} \tag{10.6}$$

where $a_1 = (\sigma_n^2)(1 + \kappa_t^2)$, $a_2 = (\sigma_n^2 / P_1)(1 + \kappa_t^2)$, $b_1 = b_2 = (\sigma_n^2 / P_r)(1 + \kappa_r^2)$, $c = \kappa_t^2 + \kappa_r^2 + \kappa_t^2 \kappa_r^2$. The individual SNRs of RF and FSO links can be defined as $\gamma_1 = \frac{P_1 |h_r|^2}{\sigma_n^2}$ and $\gamma_2 = \frac{(\eta I)^r}{\sigma_n^2}$.

10.7 Experiment

The CNN model is trained using the channel estimates on the FSO link. The input layer has four neurons which is followed by *Conv1D* layer with 64 filters. The output of this layer is provided to the *MaxPooling* layer for dimensional matching. The output layer is the *Dense* layer with 50 neurons. The performance of the CNN model for FSO channel estimation is shown in Fig. 10.2. The MSE of the CNN model has been obtained as $1.5824e - 05$ which is $MSE = -48.006$ dB. Also, the RMSE of the trained CNN model is $RMSE = -16.4$ dB.

The LSTM model has two hidden layers which join the input and the output layers. The first hidden layer consists of ten units, while the second hidden layer has five units. The loss function used for training LSTM model is the *mse*, while the *Adam* optimizer has been used. A demonstration of training and validation losses is provided in Fig. 10.3. The MSE for the trained LSTM model has been found to be 0.00078 which is −31.05 in dB, while the RMSE is −12.08 dB. Interestingly, both the models perform better when compared to the traditional ANN model. The MSE for the ANN model for channel estimation is obtained around −30 dB, and RMSE has been obtained around −5.54 dB.

For the purpose of evaluation of the performance of the models, MSE and RMSE are calculated. The FSO link dataset of irradiance values I is converted as I_{test} and I_{train}. The test data is of length N_2, while the training data is of length N_1. The MSE is defined as

$$MSE = \frac{1}{N_2} \sum_{i=1}^{N_2} (I_{test} - I_{pred})^2 \tag{10.7}$$

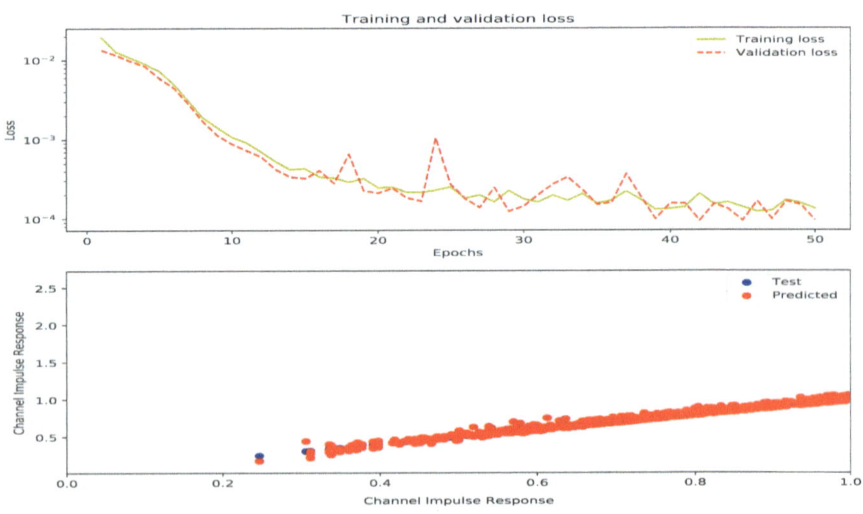

Fig. 10.2 The performance of the CNN model for the channel estimation of FSO link

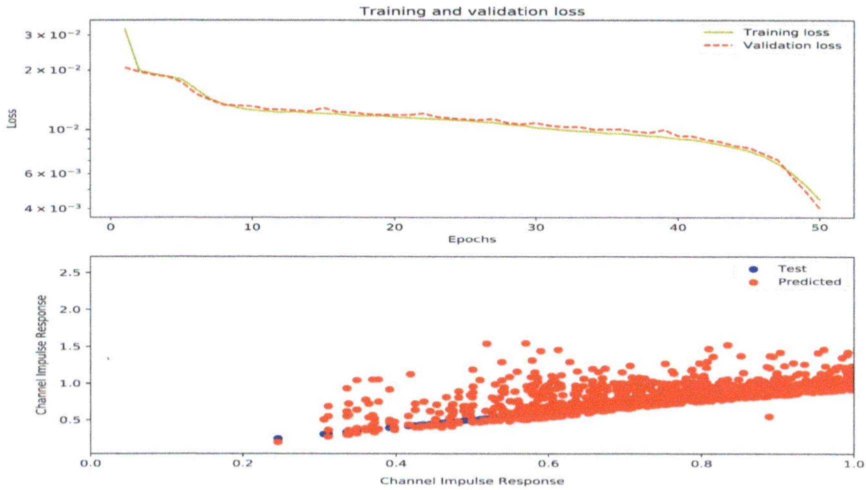

Fig. 10.3 The performance of the LSTM model for the channel estimation of FSO link

where I_{test} and I_{pred} are samples of the channel response from the test and predicted values. On the other hand, the RMSE values can be evaluated using the relation

$$RMSE = \sqrt{\frac{1}{N_2}\sum_{i=1}^{N_2}(I_{test} - I_{pred})^2} \tag{10.8}$$

When the FSO channel estimates are obtained after training, the SNR on the FSO link can be defined as $\hat{\gamma}_2 = \frac{(\eta I_{pred})^r}{\sigma_n^2}$. The corresponding SNDR values can be defined as

$$S N\hat{D} R_1 = \frac{\gamma_1 \hat{\gamma}_2}{\frac{\gamma_1^2}{P_r} c + \gamma_1 \hat{\gamma}_2 c + \hat{\gamma}_2 P_1 b_1 + \gamma_1 (a_1 + P_1 b_1) + \frac{\sigma_n^4}{P_r}} \tag{10.9}$$

$$S N\hat{D} R_2 = \frac{\gamma_1 \hat{\gamma}_2}{\frac{\hat{\gamma}_2^2}{P_r} c + \gamma_1 \hat{\gamma}_2 c + \gamma_1 b_2 + \hat{\gamma}_2 (a_2 + P_1 b_2) P_1 + \frac{\sigma_n^4}{P_1 P_r}} \tag{10.10}$$

10.8 Results

The outage probability results have been plotted in Fig. 10.4. A comparison between the outage probability values computed using the test dataset is provided with the outage probabilities calculated using the CNN and LSTM models. The performance

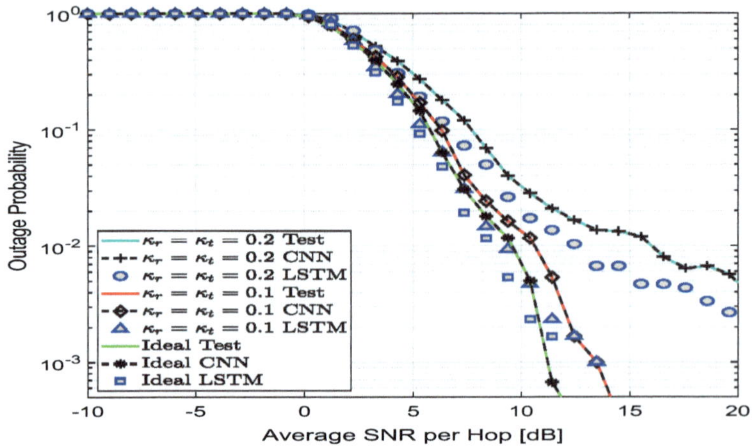

Fig. 10.4 Outage probability analysis comparing the performance based on test data with channel estimation performed using CNN and LSTM models

of the optical network is highly affected by the fast switching speeds of the hardware involved. The effect of hardware impairment in two-way RF/FSO relaying has been shown in the results. The relay node is assumed to function with hardware impairment. It has both receiveand transmit hardware impairment. These two are represented as κ_r and κ_t, respectively. The plot shows that increase in hardware impairment at the relay node increases the probability of outage for the TWR RF/FSO relay networks. Specifically, when $\kappa_r = \kappa_t = 0.1$, the outage probability is less than when $\kappa_r = \kappa_t = 0.2$. Compared to these scenarios, the performance is the best when the hardware involved is ideal. The test results are plotted on the same axis where the outage probability values calculated using the CNN and LSTM models are shown. As per the accuracy of the models, the outage probability values evaluated using the CNN model are much nearer to the test data when compared to the LSTM model.

The ergodic capacity of the hardware impaired system model is depicted in Fig. 10.5. Here, two cases are considered. First, the case when the relay node has no receive or transmit hardware impairment has been shown. Second, the case with hardware impairment with $\kappa_r = \kappa_t = 0.2$ has been considered. It can be seen in the plot that the case with ideal relay hardware provides extensively higher throughput when compared to the case when there is impairment. The TWR RF/FSO system is clearly impacted by the relay hardware impairment. The ergodic capacity has also been evaluated using the channel estimates on the FSO link. For the case of capacity calculation, the accuracy of the CNN and LSTM models in the channel estimation is comparable. This means that deep learning models can be used for channel estimation on the FSO link and can be used for calculation of performance metrics when they are trained aptly.

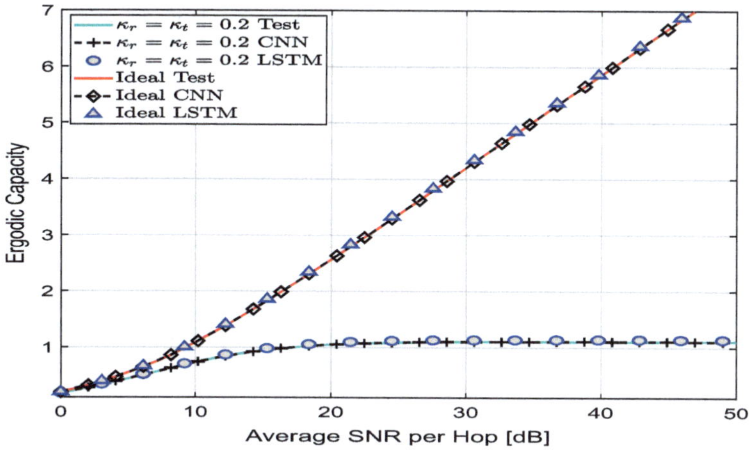

Fig. 10.5 Ergodic capacity analysis comparing the performance based on test data with channel estimation performed using CNN and LSTM models

10.9 Conclusion

The future wireless communication enablers are focusing on implementing FSO technology for high data rates. The deep learning models are being implemented for performance evaluation of the future generations of wireless technology. With an aim to implement FSO with deep neural networks, the present chapter focuses on channel estimation of optical wireless channel using CNN and LSTM techniques. For the purpose of demonstration, a TWR RF/FSO cooperative relaying technique is investigated. The channel estimation has been performed on the FSO link. The accuracy of both the deep learning models has been evaluated. The analysis has been extended to simulate the outage probability and capacity of the TWR RF/FSO relaying system. The test data is used to obtain the outage probability and capacity of the system. These outage and throughput values are compared with the performance metrics evaluated using the channel estimates using the CNN and LSTM models. It can be seen in the simulation results that it is possible to use the CNN and LSTM models for performance evaluation on the FSO systems with requisite accuracy.

References

1. P. Dong, H. Zhang, G.Y. Li, I.S. Gaspar, N.N. Alizadeh, Deep CNN-based channel estimation for mmWave massive MIMO systems. IEEE J. Sel. Top. Sign. Proces. **13**(5), 989–1000 (2019)
2. P. Jiang, C.-K. Wen, S. Jin, G.Y. Li, Dual CNN-based channel estimation for MIMO-OFDM systems. IEEE Trans. Commun. **69**(9), 5859–5872 (2021)
3. A.K. Gizzini, M. Chafii, A. Nimr, R.M. Shubair, G. Fettweis, CNN aided weighted interpolation for channel estimation in vehicular communications. IEEE Trans. Veh. Technol. **70**(12), 12796–12811 (2021)

4. P. Dong, H. Zhang, G.Y. Li, N.N. Alizadeh, I.S. Gaspar, Deep CNN for wideband mmWave massive MIMO channel estimation using frequency correlation, in *ICASSP 2019—2019 IEEE International Conference on Acoustics, Speech and Signal Processing (ICASSP)* (2019), pp. 4529–4533
5. B. Lin, X. Wang, W. Yuan, N. Wu, A novel OFDM autoencoder featuring CNN-based channel estimation for internet of vessels. IEEE Internet Things J. **7**(8), 7601–7611 (2020)
6. Y. Liao, Y. Hua, X. Dai, H. Yao, X. Yang, ChanEstNet: A deep learning based channel estimation for high-speed scenarios, in *ICC 2019—2019 IEEE International Conference on Communications (ICC)* (2019), pp. 1–6
7. Z. Wang, F. Pu, X. Yang, N. Chen, Y. Shuai, R. Yang, Online LSTM-based channel estimation for HF MIMO SC-FDE system. IEEE Access **8**, 131005–131020 (2020)
8. S. Moon, H. Kim, I. Hwang, Deep learning-based channel estimation and tracking for millimeter-wave vehicular communications. J. Commun. Networks **22**(3), 177–184 (2020)
9. T. Peng, R. Zhang, X. Cheng, L. Yang, LSTM-based channel prediction for secure massive MIMO communications under imperfect CSI, in *ICC 2020—2020 IEEE International Conference on Communications (ICC)* (2020), pp. 1–6
10. I. Goodfellow, Y. Bengio, A. Courville, *Deep Learning* (MIT Press, New York, 2016)
11. C.C. Aggarwal, *Neural Networks and Deep Learning* (Springer, Cham, 2023), 2(XXIV, 529)
12. M. Bkassiny, Y. Li, S.K. Jayaweera, A survey on machine-learning techniques in cognitive radios. IEEE Commun. Surv. Tutorials **15**(3), 1136–1159 (2013)

Index

A. Upadhya et al., *Free-Space Optical Communication Systems for Next Generation Networks*, https://doi.org/10.1007/978-3-031-74805-9

MIX
Papier aus verantwortungsvollen Quellen
Paper from responsible sources
FSC® C105338

If you have any concerns about our products,
you can contact us on
ProductSafety@springernature.com

In case Publisher is established outside the EU,
the EU authorized representative is:
Springer Nature Customer Service Center GmbH
Europaplatz 3, 69115 Heidelberg, Germany

Printed by Libri Plureos GmbH
in Hamburg, Germany